PREFACE

This manual provides technical information, training techniques, and combat techniques of light antiarmor weapons. Intended users include leaders and designated soldiers who will use this information to successfully integrate light antiarmor weapons into combat operations. This manual discusses gunnery training and train-the-trainer.

The tactical positions shown in this manual were drawn to enhance the reader's understanding of related subject material and do not necessarily represent the best employment option for that material.

Unless this publication states otherwise, masculine nouns and pronouns do not refer exclusively to men.

The proponent of this publication is the United States Army Infantry School. Send comments and recommendations on DA Form 2028 (Recommended Changes to Publications and Blank Forms) directly to—

Commandant
US Army Infantry School
ATTN: ATSH-IN-S3
Fort Benning, GA 31905-5595
lusanoh@benning.army.mil

CHAPTER 1
INTRODUCTION

This chapter provides information common to the light antiarmor weapons discussed in this manual. Topics include care and handling, destruction and decontamination procedures, and operating temperatures. Light antiarmor weapons are used against light armored vehicles, field fortifications, or other similar targets. These weapons are issued as rounds of ammunition to individual soldiers in addition to their assigned weapons and the unit's organic antiarmor weapons. Light antiarmor weapons can withstand extreme weather and environmental conditions, including arctic, tropical, and desert. The light antiarmor weapons category includes both light antiarmor and light antitank weapons.

1-1. TYPES OF LIGHT ANTIARMOR WEAPONS

Light antiarmor weapons include the M72-series light antitank weapon (LAW) and the M136 AT4. The M72-series LAW was designed in the early 1960's for use against light tanks of that era (Figure 1-1). The M136 AT4 was designed in the late 1980's for use against the improved armor of light armored vehicles (Figure 1-2).

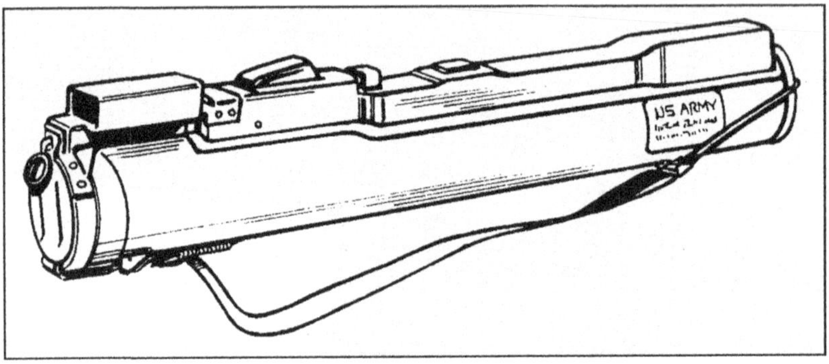

Figure 1-1. M72-series light antitank weapon.

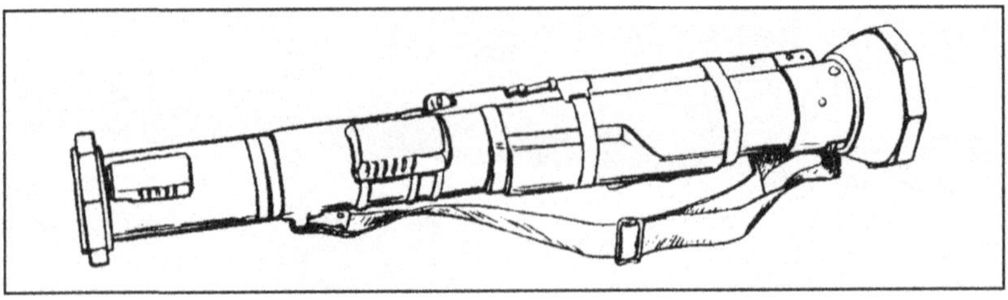

FIELD MANUAL
NO. 3-23.25 (FM 23-25)

HEADQUARTERS
DEPARTMENT OF THE ARMY
WASHINGTON, DC, 30 August 2001

LIGHT ANTIARMOR WEAPONS

CONTENTS

*This publication supersedes FM 23-25, 17 August 1994.

Page

Figure 1-2. M136 AT4 light antiarmor weapon.

1-2. CARE AND HANDLING

Light antiarmor weapons are issued as rounds of ammunition. The only requirement for their care is a visual inspection, outlined in the appropriate chapter for each weapon (Chapter 2 for the LAW and Chapter 3 for the AT4).

1-3. DESTRUCTION PROCEDURES (COMBAT ONLY)

In combat, live and expended light antiarmor weapons are destroyed only to prevent their capture or use by the enemy and, even then, only on order. For such an order to be given, the weapons must be so badly damaged that neither repairs nor cannibalization can restore them to usable condition (FM 5-25). Table 1-1 provides destruction procedures for live and expended light antiarmor weapons; Appendix A discusses safety precautions to follow when destroying them.

DANGER

TO AVOID POSSIBLE INJURY OR DEATH, MOVE TO A SAFE POSITION AND TAKE COVER BEFORE USING ANY DESTRUCTION PROCEDURE. BEFORE USING DEMOLITIONS FOR ANY REASON, YOU MUST KNOW THE PROPER PROCEDURES IN FM 5-25.

LIVE LIGHT ANTIARMOR ROUND	DEMOLITION	Prepare a 113-gram (one-quarter pound) demolition charge. Tape or tie the charge over the propellant charge. Dual prime the charge to reduce the chance of a misfire.
	BURNING	Construct a pit or trench deep enough to allow 0.6 meter (2 feet) of space between the weapons and the top surface of the ground. Place combustible material such as wood, paper, or rags in the pit, then place the weapon inside, pointed into the side of the pit and directed away from all friendly soldiers. Pour diesel fuel or oil over the weapons and the combustible material.
	FIRING	If time does not permit use of the previous methods, dispose of the weapons by firing them randomly in the direction of the enemy. Before using this method, observe all appropriate safety requirements.

Table 1-1. Destruction procedures for light antiarmor weapons.

EXPENDED LIGHT ANTIARMOR ROUND	MECHANICAL	Though you must never use mechanical means to destroy live antiarmor weapons or expended M136 AT4s, you may do so to destroy the residue from an expended M72-series LAW. For example, you can drive over it with a tracked vehicle or strike it with a pick, ax, or other object, as long as you make it unusable.
	DEMOLITION	Same as a live round.
	BURNING	Same as a live round.

Table 1-1. Destruction procedures for light antiarmor weapons (continued).

> **DANGER**
>
> **1. WHEN USING FIRE TO DESTROY A LIGHT ANTIARMOR WEAPON, THE TIME REQUIRED TO EXPLODE THE WARHEAD IS UNPREDICTABLE. ALSO, IGNITING THE PROPELLANT CAN CAUSE IT TO FIRE THE WARHEAD IN ANY DIRECTION, WHICH COULD IN TURN CAUSE INJURY OR DEATH.**
>
> **2. OBSERVE THE APPROPRIATE SAFETY PRECAUTIONS WHEN HANDLING DIESEL FUEL. CARELESSNESS COULD CAUSE PAINFUL, EVEN FATAL, BURNS.**
>
> **3. DO NOT TRY TO USE VEHICLES OR MECHANICAL MEANS TO DESTROY LIVE ANTIARMOR WEAPONS. EITHER METHOD COULD DETONATE THE WARHEAD OR PROPELLANT CHARGE, WHICH COULD CAUSE INJURY OR DEATH.**

1-4. DECONTAMINATION PROCEDURES

The soldier can use his M258A1 or DKIE (XM280) individual decontamination packet to remove H-series, G-series, and V-series agents. FM 3-5 provides more information about decontamination procedures for equipment and weapons.

> **DANGER**
>
> **NEVER USE DS2 TO DECONTAMINATE ANY LIGHT ANTIARMOR WEAPON. THE DS2 WOULD DISSOLVE THE RUBBER AND PLASTIC SEALS, ALLOWING THE DS2 TO REACH THE PROPELLANT AND PRODUCING AN EXTREMELY HAZARDOUS MIXTURE**

1-5. OPERATING TEMPERATURES

Operating temperatures for the M72-series LAW and M136 AT4 are -40 degrees to 140 degrees Fahrenheit (-40 degrees to 60 degrees Centigrade). Firing light antiarmor weapons in temperatures outside these limits could cause a misfire or produce some other hazard for the soldier (Appendix A).

CHAPTER 2

M72-SERIES LAW, OPERATION AND FUNCTION

This chapter provides information on and technical data for the M72-series light antitank weapon (LAW). It also discusses the characteristics, nomenclature, functioning, and operation of the LAW.

2-1. DESCRIPTION

The M72-series LAW is a lightweight, self-contained, antiarmor weapon consisting of a rocket packed in a launcher (Figure 2-1). It is man-portable, may be fired from either shoulder, and is issued as a round of ammunition. It requires little from the user—only a visual inspection and some operator maintenance. The launcher, which consists of two tubes, one inside the other, serves as a watertight packing container for the rocket and houses a percussion-type firing mechanism that activates the rocket.

a. **Outer Tube**. The trigger housing assembly (which contains the trigger assembly) is on the upper surface of the outer tube. So are the trigger arming handle, front and rear sight assemblies, and the launcher's rear cover.

b. **Inner Tube**. The inner tube telescopes outward toward the rear, guided by a channel assembly that rides in an alignment slot in the outer tube's trigger housing assembly. The channel assembly also houses the firing pin rod assembly, which includes a detent lever assembly. The detent lever assembly moves under the trigger assembly in the outer tube, locking the inner tube in the extended position and cocking the weapon. All this must occur before the weapon can be fired.

c. **Rocket**. The rocket is a percussion-ignited, fin-stabilized, fixed munition. The rocket is attached by the igniter to the inside of the launcher. The rocket consists of a 66-mm HEAT warhead, a point-initiating, base-detonating fuze, and a rocket motor. Six spring-loaded fins are attached to the rear of the rocket motor. These fins are folded forward along the motor when the rocket is in the launcher. When ignited, the propellant in the rocket motor burns completely, producing gasses about 1,400 degrees Fahrenheit (760 degrees Centigrade). The gas pressure pushes the rocket toward the target and exits to the rear of the launcher as the backblast.

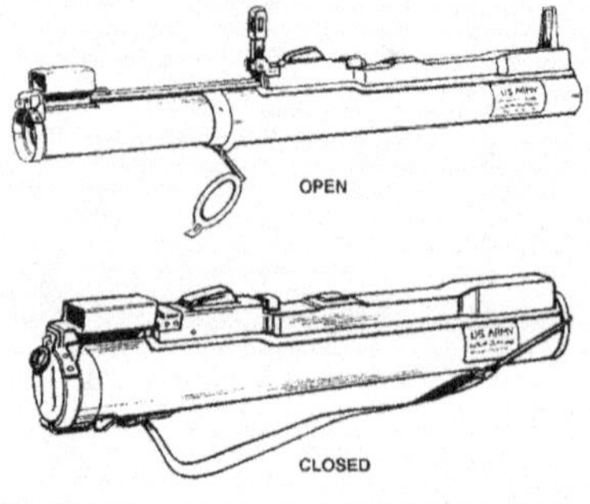

Figure 2-1. The M72A2 or M72A3 LAW.

2-2. TECHNICAL DATA

The following data apply to the M72A2 and M72A3 LAWs:

a. **Launcher**.

Length (Extended) .... Less than 1 meter (34.67 inches)
Length (Closed) .... 0.67 meter (24.8 inches)
Weight (Complete M72A2) .... 2.3 kg (5.1 pounds)
Weight (Complete M72A3) .... 2.5 kg (5.5 pounds)
Firing Mechanism .... Percussion
Front Sight .... Reticle graduated in 25-meter range increments
Rear Sight .... Peep sight adjusts automatically to temperature change

b. **Rocket**.

Caliber .... 66 mm
Length .... 50.8 cm (20 inches)
Weight .... 1.8 kg (2.2 pounds)
Muzzle Velocity .... 144.8 mps (475 fps)
Minimum Range (Combat) .... 10 meters (33 feet)
Minimum Arming Range .... 10 meters (33 feet)
Maximum Range .... 1,000 meters (3,300 feet)
Maximum Effective Ranges:
Stationary Target .... 200 meters (660 feet)
Moving Target .... 165 meters (541 feet)
(Beyond these ranges, there is less than a fifty percent chance of hitting the target.)

2-3. AMMUNITION

The M72-series LAW is issued as a round of ammunition. It contains a nonadjustable propelling charge and a rocket. Every M72-series LAW has an integral high-explosive antitank (HEAT) warhead. The warhead is in the rocket's head (or body) section. The fuze and booster are in the rocket's closure section. The propellant, its igniter, and the fin assembly are in the rocket's motor. No inert versions are available (Figure 2-2). Appendix B provides information about appropriate gunnery training devices and ammunition. Although the M72-series LAW is mainly used as an antiarmor weapon, it may be used with limited success against secondary targets such as gun emplacements, pillboxes, buildings, or light vehicles. (Chapter 6 provides more information about combat techniques.)

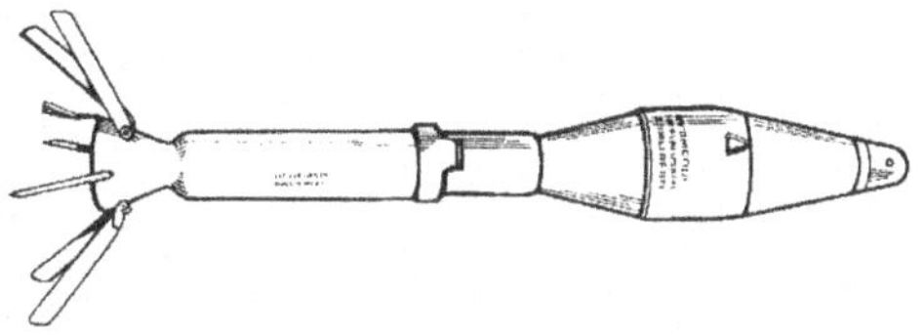

Figure 2-2. High-explosive antitank rocket (ammunition).

a. **Description**. The 66-mm HEAT rocket warhead consists of a tapered, thin-gauge steel body. When it explodes, the force and heat of the explosive focus into a small but powerful gas jet. This directional jet penetrates the target and, if the target is a vehicle, sprays molten metal inside. If the jet hits an engine or ammunition, it may start a fire or cause an explosion. Figure 2-3 shows how the warhead penetrates 300 millimeters of rolled homogeneous steel armor.

(1) ***Impact***. The nose cone crushes; the impact sensor activates the fuze.

(2) ***Ignition***. The ogive crush switch activates the electric detonator. The booster detonates, initiating the main charge.

(3) ***Penetration***. The main charge fires and forces the warhead body liner into a directional gas jet that penetrates armor plate.

(4) ***After Armor Effects (Spalling)***. The projectile fragments and incendiary effects produce blinding light and destroy the target's interior.

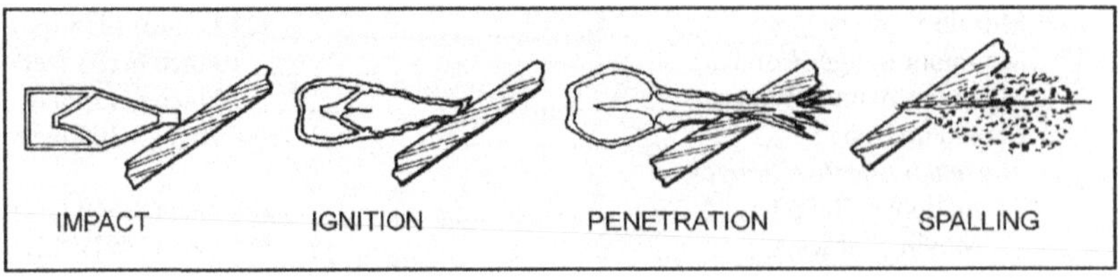

Figure 2-3. Effects of M72-series LAW warhead.

b. **Characteristics**. The head of the round is olive drab stenciled in yellow. The M412 fuze is dropsafe and boresafe. Its minimum arming distance is about 33 feet (10 meters). Six stabilizing fins are attached as part of the motor. As the rocket clears the launcher, springs force open the fins, which stabilize the rocket in flight.

c. **Packaging**. Five complete M72-series LAWs are packaged within a fiberboard inner pack for a total weight of 12.5 kilograms (27 1/2 pounds). Three inner packs are then placed in a wire-bound wooden box, the gross weight of which is 54.5 kilograms (120 pounds) (Figure 2-4, page 2-4).

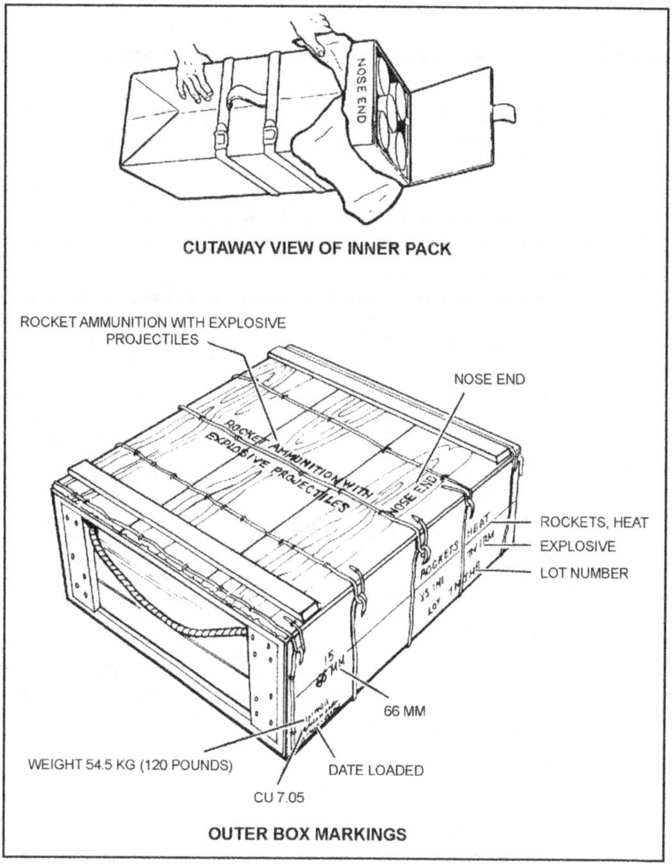

Figure 2-4. Packaging for M72-series LAW.

2-4. INSPECTION

Because the M72-series LAW is issued as a round of ammunition rather than as a weapon, inspection is limited to a visual examination of the sealed unit.

Inspect the launcher's overall condition before preparing the launcher for use.

- Check the body for dents, cracks, or bulges.
- Check the rubber boots covering the trigger bar and barrel detent for tears or punctures.
- Ensure the arming handle is present and on SAFE and that the pull pin is in place.

- Check the data plate for the phrase, W/COUPLER (Figure 2-5).

DANGER

1. IF THE M72A2 LAW DOES NOT STATE "W/COUPLER" ON ITS DATA PLATE, TURN THE WEAPON IN TO THE UNIT AMMUNITION SECTION.

2. THE COUPLER PREVENTS THE INNER AND OUTER TUBES FROM SEPARATING AND POSSIBLY CAUSING PREMATURE DETONATION.

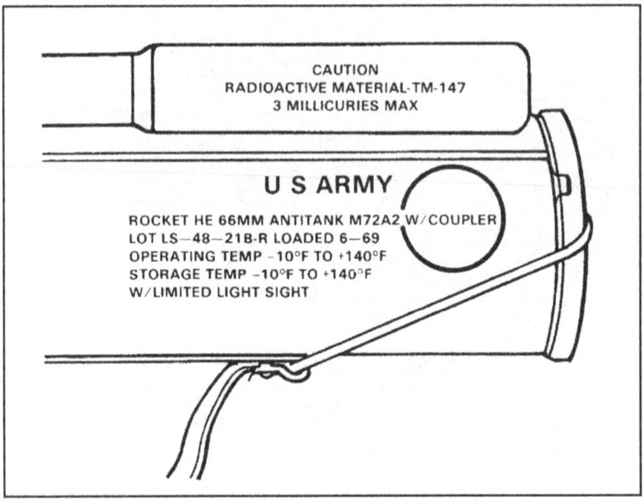

Figure 2-5. Launcher data plate.

2-5. FIRING MECHANISM

The firing mechanism includes the trigger arming handle, the trigger assembly, and the firing pin rod assembly (Figure 2-6).

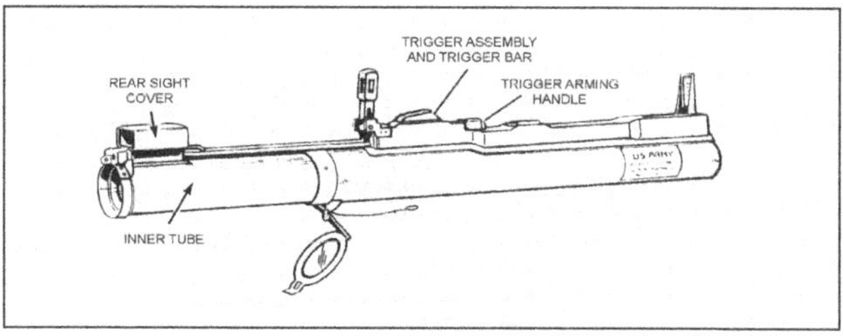

Figure 2-6. Firing mechanism.

a. **Trigger Arming Handle**. The trigger arming handle is located forward of the trigger bar and has two positions: SAFE and ARM. Leave the trigger arming handle on SAFE until the launcher is in the correct firing position (Figure 2-7). To press the trigger, you must first pull the arming handle forward and lock it in the ARM position.

b. **Trigger Assembly**. The trigger assembly is on the top rear of the outer tube. To fire the launcher, press downward on the trigger bar.

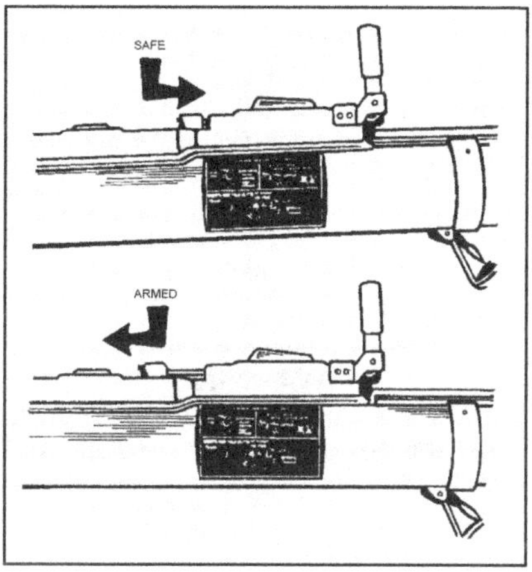

Figure 2-7. Trigger arming handle.

c. **Firing Pin Rod Assembly**. The rear sight cover and the firing pin housing are on the top of the rear of the inner tube. Inside the housing, the primer and the firing pin rod are aligned (Figure 2-8). Pressing the trigger bar releases the tension on the firing pin rod assembly, allowing the firing pin to strike the center of the primer.

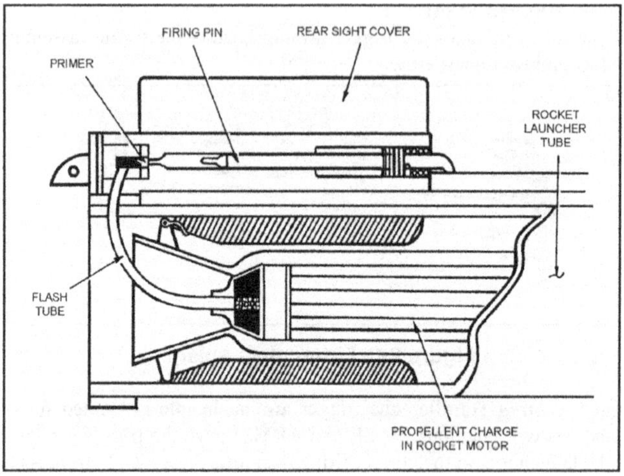

Figure 2-8. Firing pin, primer, and propellant charge.

2-6. SIGHTS

This paragraph discusses the front and rear sights and their proper use.

a. **Front Sight, M72A2 and M72A3 LAWs.** The front sight has a raised vertical range line marked with ranges from 50 to 350 meters in 25-meter increments (Figure 2-9). Two curved stadia lines are etched on the front sights. Do not use the stadia lines on this sight to estimate range, because they are inaccurate. Lead indicators are located on either side of the stadia lines to help you engage moving targets. On the M72A3, use the front sight illuminated range marks at the 100-meter and 150-meter points to help you engage targets in low light.

WARNING
DO NOT TOUCH THE RANGE MARKS; THEY ARE ILLUMINATED WITH PROMETHIUM, WHICH IS MILDLY RADIOACTIVE.

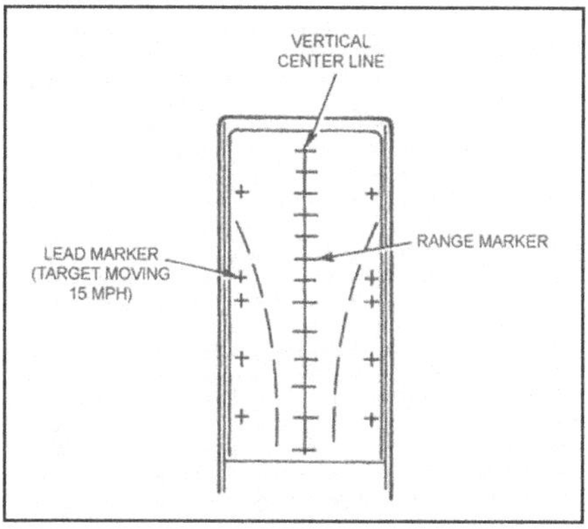

Figure 2-9. Front sight, M72A2 and M72A3 LAWs.

b. **Rear Sight, M72A2 and M72A3 LAWs**. The rear sight consists of a steel bracket with a rubber boot and plastic peep sight. This sight automatically adjusts to changes in temperature (Figure 2-10), which means that its settings are unaffected by temperature.

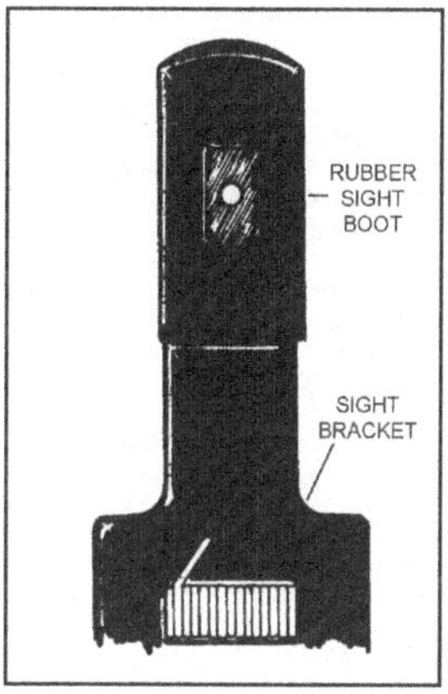

Figure 2-10. Rear sight, M72A2 and M72A3 LAWs.

c. **AN/PVS-4 Nightsight**. This nightsight is issued with various accessories, including a bracket that, when mounted on an M72-series LAW, will allow you to use an AN/PVS-4 on the LAW. However, this works only if DS maintenance has already installed an M72A1 reticle in the AN/PVS-4. Though the reticle was developed for the M72A1 LAW, you can also use it with the other M72-series models. To mount the M72A1 bracket assembly on any M72-series LAWC

(1) Place the bracket assembly on top of the rocket launcher (Figure 2-11) so that the square cutout in the top of the bracket fits over the extension release button.

(2) Swing the lower adapter section up and under the rocket launcher and secure it by turning the locking latch clockwise to fully engage the latch shoulder screw.

(3) Place the sight in the groove on the bracket and align the threaded screw hole in the base of the sight with the lever screw assembly. Tighten the lever screw assembly firmly.

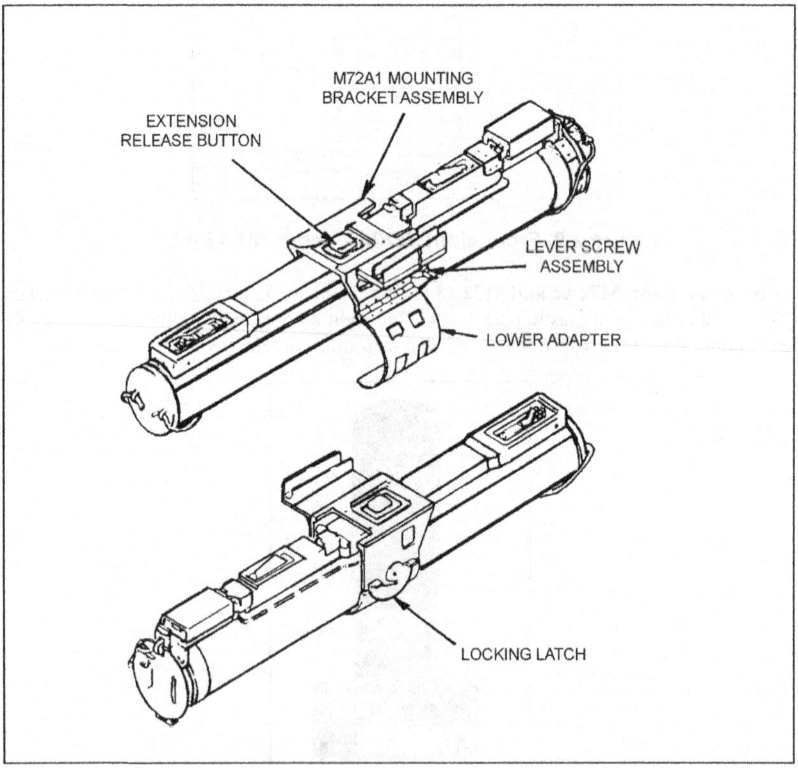

Figure 2-11. Installing M72A1 mounting bracket assembly.

(4) Use the M72A1 reticle to estimate the range to a 20-foot target such as a tank viewed from the side (Figure 2-12). Place the sight so that the rear and front of a flanking target fit between the curved vertical lines (Figure 2-13). When you read the range from the scale, note that the width of the tank is about one-half the tank's length. This means you can estimate the range to a target, whether it is headed straight toward you or straight away from you, by placing its sides between the curved vertical lines and halving the range shown at that point. The bottom of each vertical line in the center of the reticle

corresponds to an additional 25 meters. The length of a horizontal line represents 5 mils, and the spaces between lines represent 5 mils. Use the stadia lines to estimate range *only* when using the AN/PVS-4's M72A1 reticle pattern.

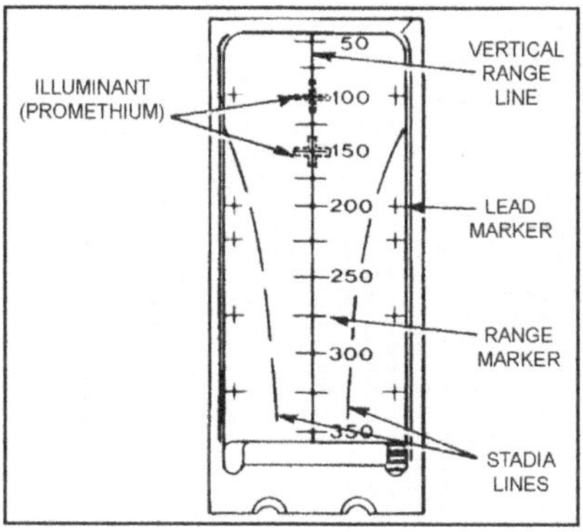

Figure 2-12. M72A1 reticle.

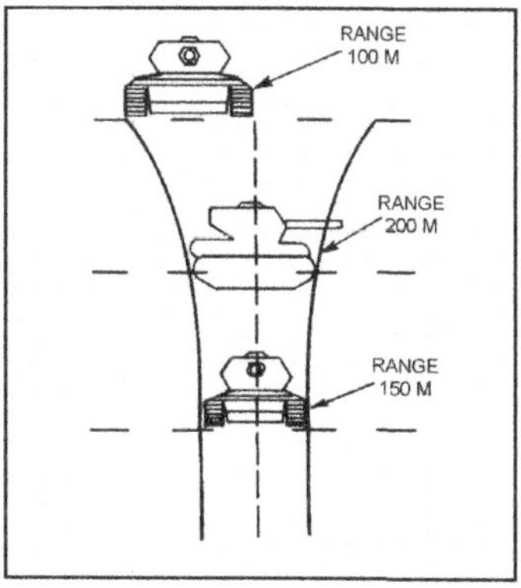

Figure 2-13. Using M72A1 stadia lines to engage targets.

(5) Align the nightsight's reticle pattern to the LAW sights only once. After that, you can move the nightsight from LAW to LAW without having to realign it each time (Figure 2-14). To align the sight reticle pattern to the LAW sights, place a target at 25 meters. Install the mounting bracket and sight, and select a stable firing position for the LAW. Turn both the TUBE BRIGHTNESS and RETICLE BRIGHTNESS knobs ON. Align the 200-meter range mark on the daysight with the aiming point on the target. Without moving the LAW, adjust the nightsight reticle so that its 200-meter range mark coincides with an aiming point 6.3 centimeters left and 3.5 centimeters above the target aiming point.

DANGER

IF YOU CANNOT FIND AN EMPTY LAUNCHER TO USE TO ALIGN THE NIGHTSIGHT, YOU MAY USE AN UNFIRED LAUNCHER, BUT THIS IS EXTREMELY DANGEROUS.

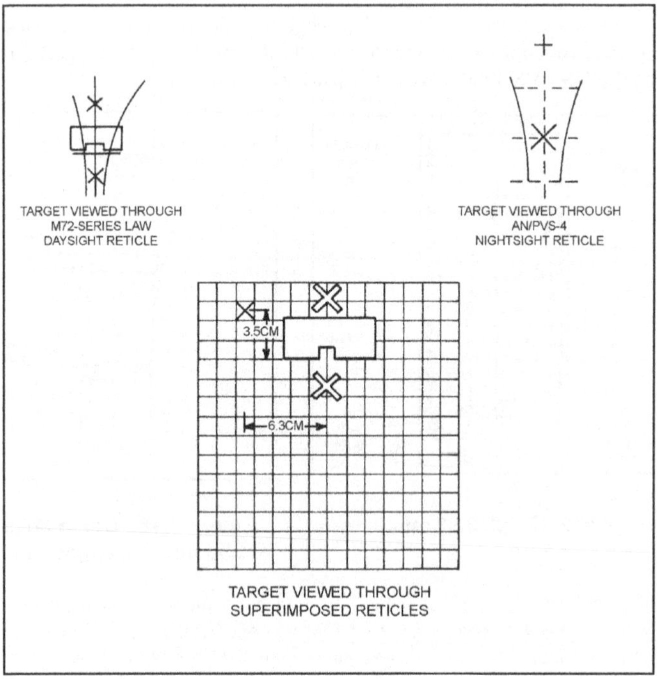

Figure 2-14. M72A1 alignment target for the AN/PVS-4.

2-7. OPERATION AND FUNCTION

Before preparing the launcher for use, the firer inspects its overall condition.

a. To extend the rocket launcher—

(1) Remove the pull pin and rotate the rear cover downward so the front cover and adjustable sling assembly can fall free (Figures 2-15 and 2-16, page 2-11). Do not discard the sling assembly until after you fire the rocket.

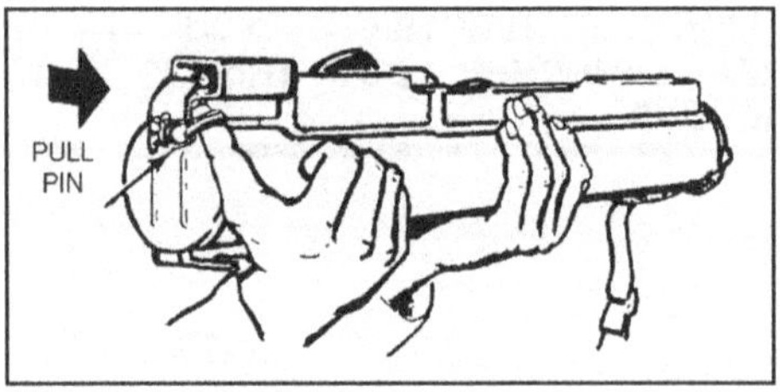

Figure 2-15. Removing the pull pin.

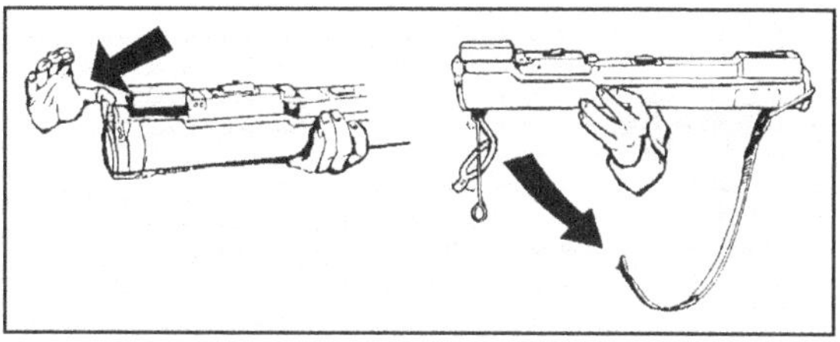

Figure 2-16. Removing the front cover and the adjustable sling assembly.

(2) With your firing hand, grasp the rear sight cover; with your nonfiring hand, grasp the launcher forward of the barrel detent. Pull your hands sharply in opposite directions to extend the launcher (Figure 2-17). To ensure the launcher is fully extended and locked, try to close it (Figure 2-18).

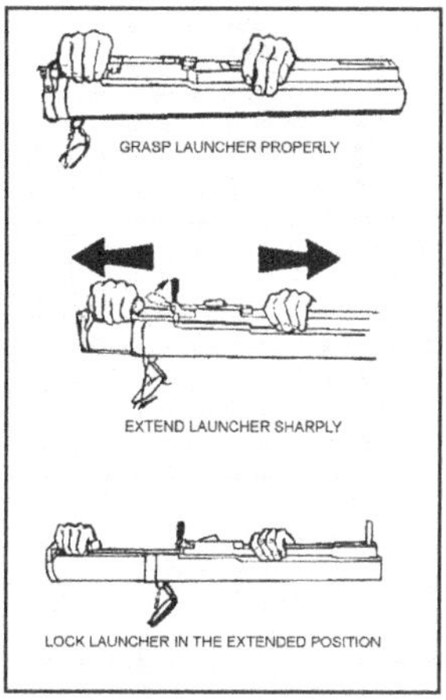

Figure 2-17. Extending the launcher.

b. To fire the rocket launcher, raise it slightly above shoulder level, rotate your body under it, and place it on your shoulder. Check the backblast area, pull the trigger arming handle to the ARM position, aim the launcher, and depress the rubber boot on the trigger bar firmly to ensure the launcher fires (Figure 2-19). (If the trigger arming handle will not remain in the ARM position, the launcher is not fully extended.) Pressing the trigger bar causes the firing pin to strike the primer, which ignites the black powder in the flash tube, which in turn ignites the propellant in the rocket motor.

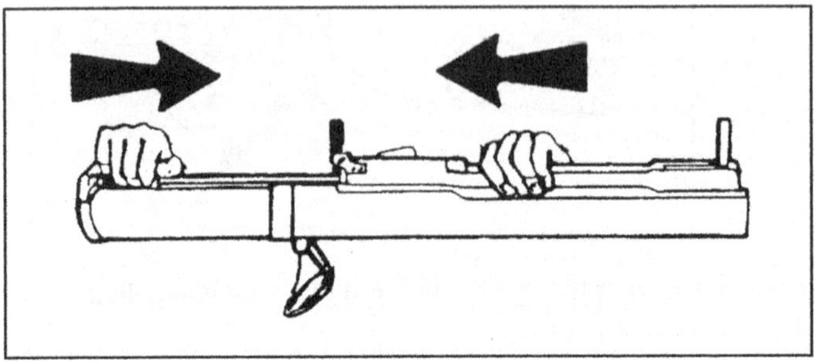

Figure 2-18. Ensuring launcher is locked in the extended position.

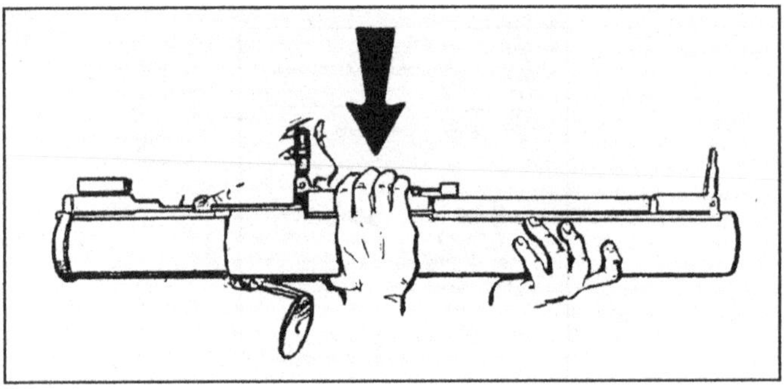

Figure 2-19. Firing the launcher.

DANGER

WHEN OPERATING THE LAW, KEEP IT POINTED DOWNRANGE. ENSURE YOUR WHOLE BODY IS CLEAR OF THE MUZZLE AND REAR OF THE LAUNCHER, AND ENSURE THE BACKBLAST AREA IS CLEAR.

2-8. MISFIRE PROCEDURES

A misfire is a complete failure to fire caused by a procedural or mechanical failure. Which misfire procedures should be used depends on whether the firer is in a combat or training environment.

a. **Causes**. A misfire is usually caused by one of the following factors:

- The launcher may not be fully extended.

- The trigger arming handle may not be armed.
- The firing mechanism or the propelling charge explosive train may be faulty.

WARNING
KEEP YOUR WEAPON POINTED TOWARD THE TARGET.

b. **Combat Environment**. If a misfire occurs in combat—

(1) Squeeze the trigger again immediately.

(2) If the launcher still fails to fire, place the trigger arming handle on SAFE.

(3) Partly collapse the launcher, then extend it to cock it again.

(4) Place it on your shoulder, check the backblast area again, then arm, aim, and fire.

(5) If the LAW still fails to fire, squeeze the trigger again and return the trigger arming handle to SAFE. Collapse the launcher, set it aside, and try another one. As soon as you can, dispose of the misfired LAW in accordance with SOP.

c. **Training Environment**. If an M72A2, M72A3, or M190 subcaliber device misfires on a live-fire training range—

(1) Squeeze the trigger again.

(2) If the launcher still fails to fire, keep the launcher on your shoulder, announce "Misfire," and wait 10 seconds. Place the trigger arming handle on SAFE.

(3) Move the launcher from your shoulder and wait one minute.

(4) Extend the launcher to cock it again, check the backblast area, place the launcher back on your shoulder, pull the arming handle to the ARM position, aim, and squeeze the trigger bar.

(5) If the launcher again fails to fire, wait 10 seconds before returning the trigger arming handle to the SAFE position.

(6) Keep the launcher trained on the target area at least one minute; DO NOT collapse the launcher.

(7) Move the launcher to a safe area and dispose of it IAW unit SOP.

2-9. RESTORATION TO CARRYING CONFIGURATION

If the launcher is prepared to fire, but then is not fired, it should be returned to the carrying configuration by reversing the preparation procedure. After the launcher has been prepared for firing, it is no longer watertight. Therefore, when carrying the launcher, sling it over either shoulder with the muzzle (forward) end down. Only the rocket and rocket motor ignition system are waterproof.

a. Return the trigger arming handle to the SAFE position.

b. Remove the launcher from your shoulder, depress the barrel detent, collapse the launcher tube, and guide the front and rear sights into position.

c. Close the rear cover, replace the cover pull pin, and replace the sling assembly.

WARNING
TO PREVENT INJURY, REMOVE YOUR THUMB FROM THE DETENT AFTER COLLAPSING THE LAUNCHER 1/2 TO 1 INCH.

CHAPTER 3

M136 AT4, OPERATION AND FUNCTION

This chapter provides information and technical data for the M136 AT4 light antiarmor weapon, including its characteristics, nomenclature, and operation. Its function, firing mechanism, and safeties are also discussed.

3-1. DESCRIPTION

The M136 AT4 is a lightweight, self-contained, antiarmor weapon. It consists of a free-flight, fin-stabilized, rocket-type cartridge packed in an expendable, one-piece, fiberglass-wrapped tube (Figure 3-1). The M136 AT4 is man-portable and is fired from the right shoulder only. The launcher is watertight for ease of transportation and storage. Though the M136 AT4 can be employed in limited visibility, the firer must be able to see and identify the target and estimate the range to it. Unlike the M72-series LAW, the M136 AT4 launcher need not be extended before firing.

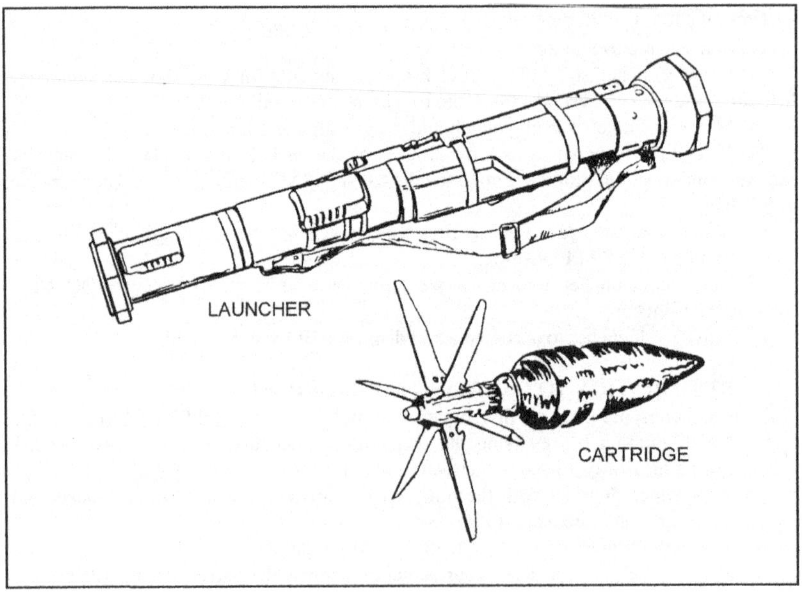

Figure 3-1. Launcher and HEAT cartridge.

3-2. TECHNICAL DATA

The following data apply to the M136 AT4:

a. **Launcher.**

Length.. 1,020 mm (40 inches)
Weight (Complete System) ....6.7 kg (14.8 pounds)
Rear Sight................................Range indicator, graduated in 50-meter increments

b. **Rocket.**

Caliber..84 mm
Muzzle Velocity ....290 mps (950 fps)
Length..460 mm (18 inches)
Weight .... 1.8 kg (4 pounds)
Minimum Range
Training ....30 meters (100 feet)
Combat ....10 meters (33 feet)
Arming ....10 meters (33 feet)
Maximum Range...2,100 meters (6,890 feet)
Maximum Effective Range ....300 meters (985 feet)

3-3. AMMUNITION

The M136 AT4 is a round of ammunition with an integral, rocket-type cartridge. The cartridge consists of a fin assembly with tracer element; a point-initiating, base-detonating, piezoelectric fuze; a warhead body with liner; and a precision-shaped explosive charge (Figure 3-2).

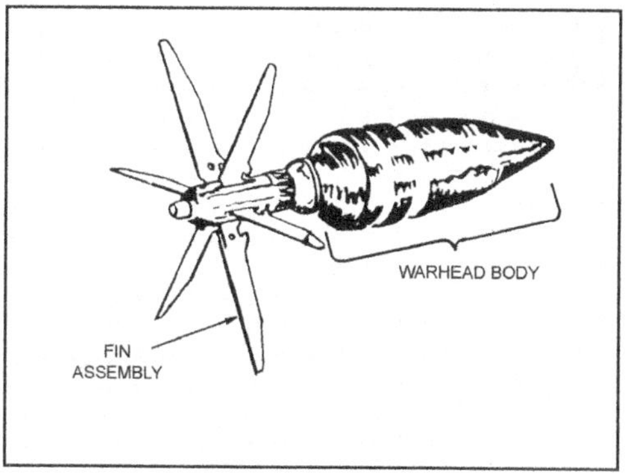

Figure 3-2. 84-mm HEAT cartridge.

a. **Description**. The M136 AT4's warhead has excellent penetration ability and lethal after-armor effects. The extremely destructive, 440 gram shaped-charge explosive penetrates more than 14 inches (35.6 cm) of armor. Warhead effects are shown in Figure 3-3.

(1) ***Impact***. The nose cone crushes; the impact sensor activates the fuze.

(2) ***Ignition***. The piezoelectric fuze element activates the electric detonator. The booster detonates, initiating the main charge.

(3) ***Penetration***. The main charge fires and forces the warhead body liner into a **directional gas jet that penetrates armor plate.**

(4) ***After-Armor Effects (Spalling)***: The projectile fragments and incendiary effects produce blinding light and destroy the interior of the target.

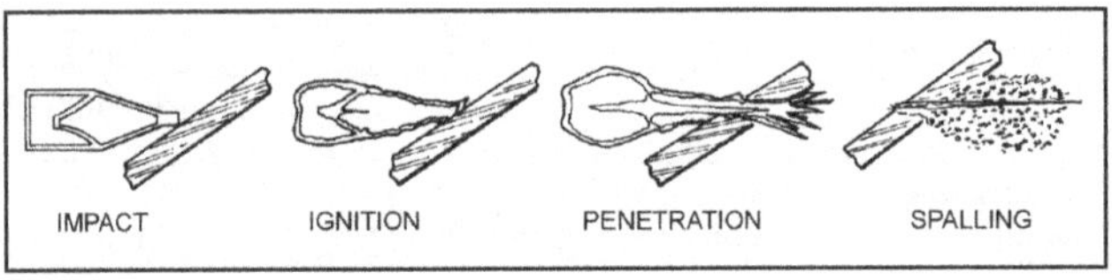

Figure 3-3. Effects of M136 AT4 warhead.

b. **Packaging**. Five M136 AT4s, each wrapped in a plastic barrier bag, are packed together in a wooden container. The containers are too heavy to stack more than four deep on the pallets (Figure 3-4).

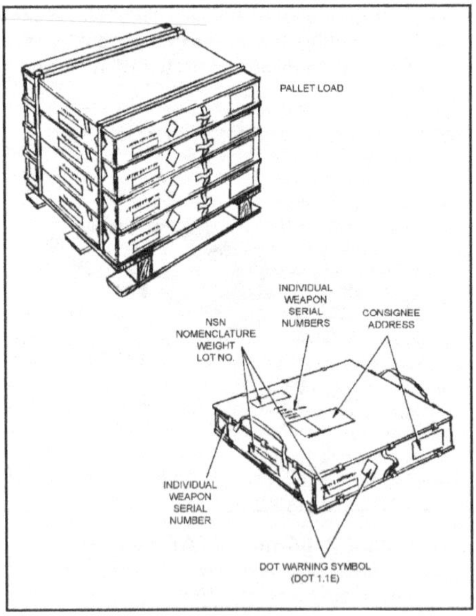

Figure 3-4. Ammunition packaging and markings.

c. **Color-Coding**. M136 AT4 launchers are marked with color-coded bands (Figure 3-5). A black with yellow band indicates an HE antiarmor round (early models had a solid black band). A gold or yellow band indicates a field handling trainer; no band indicates an M287 9-mm tracer bullet trainer (Appendix B).

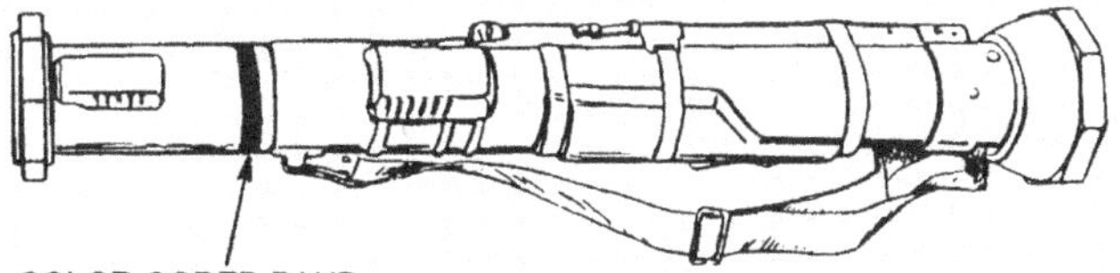

Figure 3-5. Location of color-coded band.

3-4. INSPECTION

Since the M136 AT4 is issued as a round of ammunition rather than as a weapon, the launcher is completely sealed. However, its overall condition should be inspected at the time of issue and again before use (Figure 3-6). The wooden container should be opened, the plastic bags removed, and the launcher visually inspected for obvious damage. If the M136 AT4 is not to be used immediately, it should be returned to its plastic bag and the bag resealed with tape. The soldier issued the weapon must ensure—

- The *rear seal*, a brown acrylic plastic plate inside the *venturi*, is in place and undamaged.
- The *transport safety pin* is in place and fully inserted. The lanyard is attached to the transport safety pin and the launcher. The lanyard should already be wrapped around the launcher clockwise and the transport safety pin inserted in the retainer hole counterclockwise.
- The *cocking lever* is present and in the SAFE (uncocked) position.
- The plastic *fire-through muzzle cover* is in place and undamaged. If it is torn or broken, cut it out and check the launch tube to ensure it is clear of foreign objects. Remove any that you find by turning the tube muzzle downward and gently shaking the launcher.
- The launcher has the correct *color-coded band.*
- The *sights* function properly. Open the sight covers to ensure the sights pop up and are undamaged.
- The *forward safety* does not move when you depress it.
- The *red trigger button* is not missing.
- The launcher body has no cracks, dents, or bulges.
- The *carrying sling* is not frayed and is attached firmly to the launch tube.
- The *shoulder stop* is not broken or damaged, and it unsnaps and folds down.

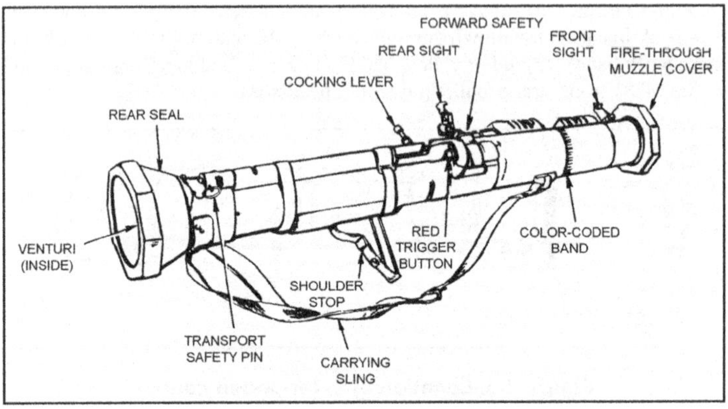

Figure 3-6. Inspection.

3-5. FIRING MECHANISM, SAFETIES, AND WEAPON FUNCTION

The function of the M136 AT4 must be discussed along with its firing mechanism and safety features. The firing mechanism is mechanical and consists of a red trigger button, an enclosed firing rod and spring, and three safety devices (Figure 3-7). The first safety device is the *transport safety* pin. The red trigger button is located between the other two, the *cocking lever* and the *forward safety*. The weapon cannot be fired until all three safeties have been disengaged.

a. **Transport Safety Pin**. This pin blocks the firing pin from striking the cartridge percussion cap. To disengage this pin, pull it outward, then release it.

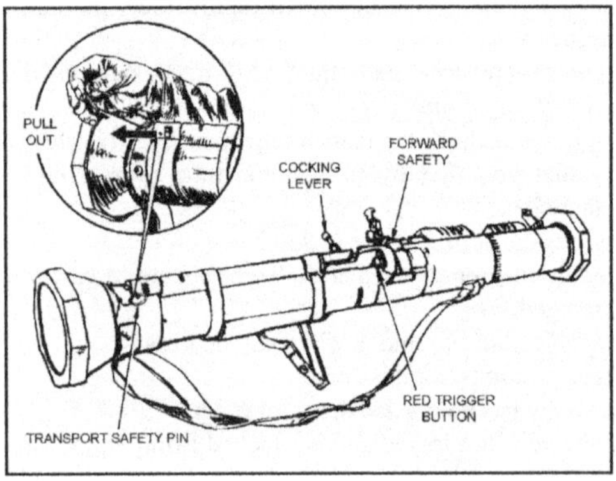

Figure 3-7. Firing mechanism and transport safety pin.

b. **Cocking Lever**. When this lever, which is attached to the firing rod (Figure 3-8), is in the SAFE position, the firing rod and the trigger cannot touch. To cock the AT4, push the lever forward and rotate it downward and to the right with your right thumb. This causes the hooks on the front of the firing rod to catch and hold the red trigger button.

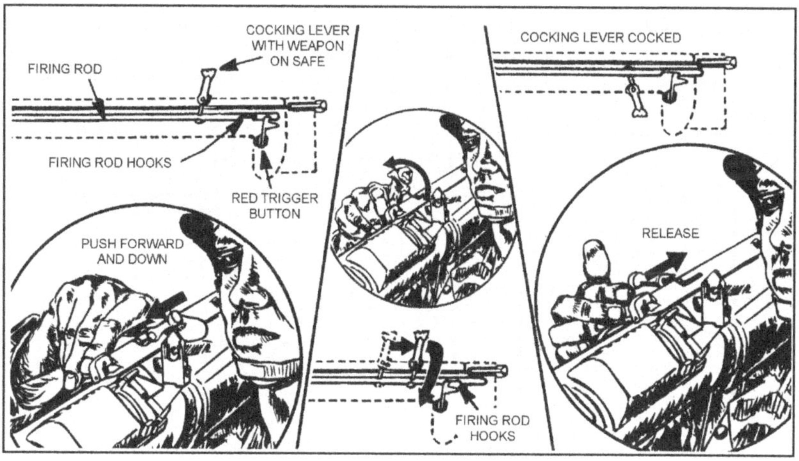

Figure 3-8. Cocking lever.

c. **Forward Safety**. This safety is on the front end of the firing mechanism (Figure 3-9) and is connected to a steel rod with a bent end that blocks the firing rod from striking the firing pin. To fire the M136 AT4, hold down the forward safety so the firing rod can strike the percussion cap and ignite the propellant when you push the trigger.

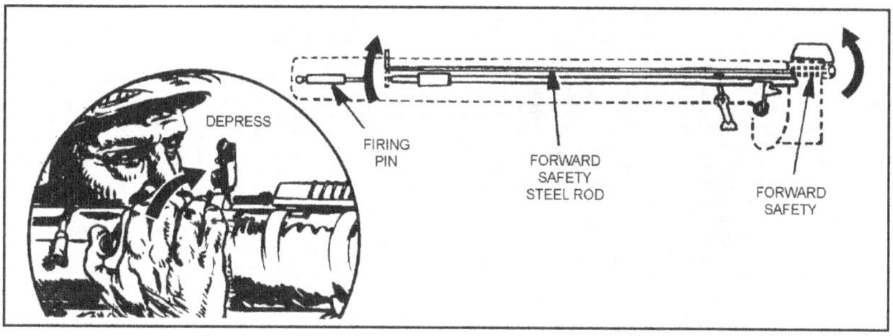

Figure 3-9. Forward safety.

3-6. SIGHTS

The fact that the AT4's front and rear sights resemble those of the M16-series rifle makes using the AT4 easier (Figure 3-10).

a. **Front Sight**. The front sight has a sight blade with a center post and left and right lead posts. A semicircular white line helps you obtain the proper sight picture. To open the front sight cover, press down on it and slide it backward until the sight pops up.

b. **Rear Sight**. The rear sight has a sight blade, range adjustment knob, range scale, 2-mm peephole for normal daylight visibility conditions, and 7-mm peephole for limited visibility conditions. To open the rear sight cover, press down on it and slide it forward until the sight pops up.

c. **Nightsight**. The AT4 can be fitted with the AN/PAQ-4C, AN/PEQ-2, or the AN/PAS-13, when used with the nightsight mounting bracket (NSN 5340-01-391-3004).

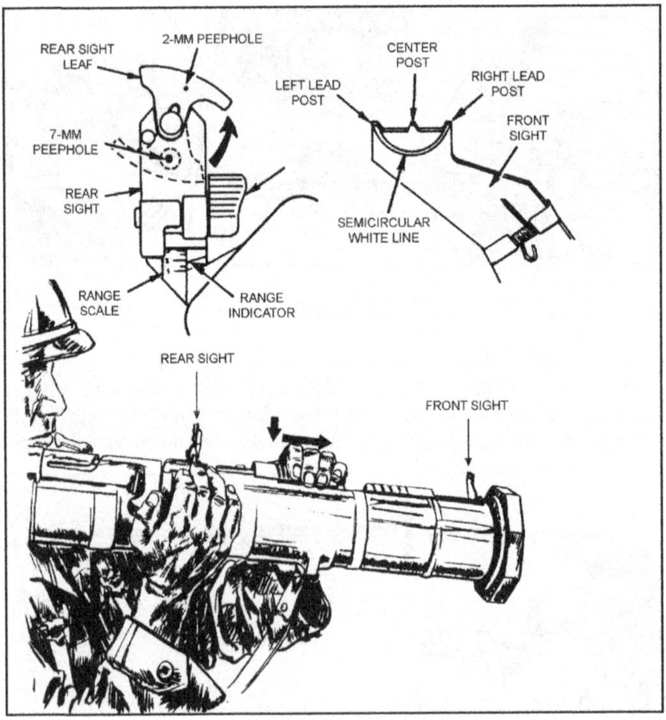

Figure 3-10. Sights.

(1) The leaf blade that covers the 7-mm peephole has its own tiny 2-mm peephole. To uncover the 7-mm peephole, pull the bottom of the leaf blade out slightly and rotate it right and up. To cover the 7-mm peephole, rotate it back down and ensure the leaf blade

is seated. The range indicator scale is indexed from 100 to 500 meters in 50-meter increments.

(2) To increase the range setting beyond 200 meters, turn the range adjustment knob clockwise, or vice versa (Figure 3-11). You must remember to reset the range to 200 meters when you close the rear sight. Otherwise, closing the sight cover will break off the rear sight.

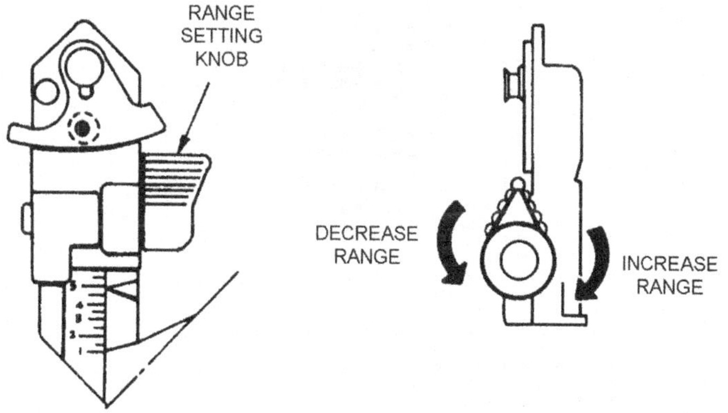

Figure 3-11. Adjusting the rear sight range setting.

3-7. OPERATION

If you are under fire, take cover before preparing the M136 AT4 for firing as follows:

a. Remove the AT4 from its carrying position and cradle it in your left arm (Figure 3-12).

> **WARNING**
> **Insert the approved brand of earplugs before you fire. Keep the weapon pointed toward the target, and keep the backblast area clear.**

Figure 3-12. Cradle position.

b. With your right hand, pull and release the transport safety pin (Figure 3-13). This pin is important—you must reinsert it if you do not fire the launcher. Therefore, unless it is attached to the launcher with a lanyard, you must keep it in a safe place.

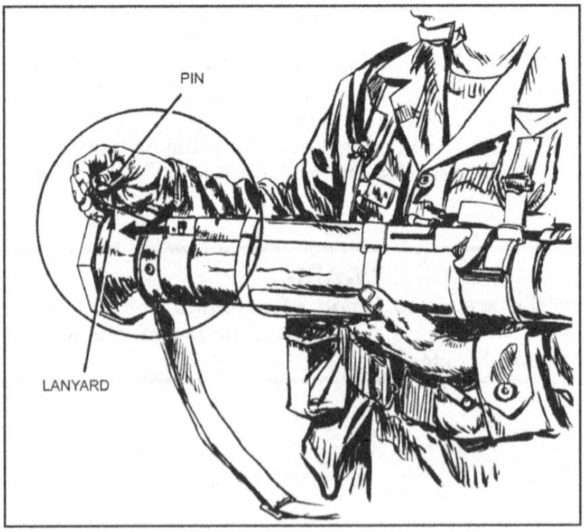

Figure 3-13. Removing the transport safety pin.

c. Unsnap, unfold, and hold the shoulder stop with your right hand (Figure 3-14).

d. Place the launcher on your right shoulder and stabilize it by grasping the sling near the launcher's muzzle with your left hand.

Figure 3-14. Unsnapping the shoulder stop.

e. With the AT4 on your right shoulder, stabilize it with your left hand and open the sights with your right hand. Press down and pull backward on the front sight cover until the front sight pops up (Figure 3-15), then press down and forward on the rear sight cover until the rear sight pops up. The rear sight should be no less than 2 1/2 inches and no more than 3 inches from your eyes.

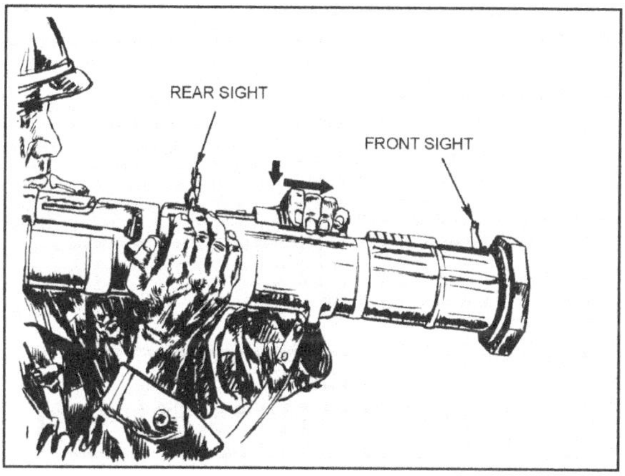

Figure 3-15. Opening and adjusting the sights.

f. Set the rear sight for the correct range to the target.

g. Check the backblast area before you cock the launcher. Then, with your right hand, unfold the cocking lever (Figure 3-16). Place your thumb under it and, with the support of your fingers in front of the firing mechanism, push it forward, rotate it downward and to the right, and let it slide backward.

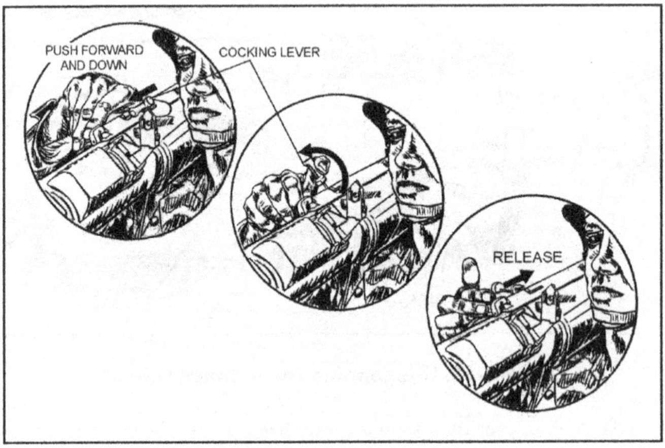

Figure 3-16. Cocking the launcher.

h. Pull back on the sling with your left hand to seat the shoulder stop firmly against your shoulder. To avoid a misfire, use the index and middle fingers on your right hand to hold the forward safety down and to the left while you fire (Figure 3-17, page 3-12).

3-8. MISFIRE PROCEDURES

A misfire is a complete failure to fire caused by a procedural or mechanical failure. Which misfire procedures should be used depends on whether the firer is in a combat or training environment.

a. **Causes.** A misfire is usually caused by one of the following factors:

The forward safety is not depressed far enough to disengage the safety.

The firing mechanism is faulty.

The propelling charge explosive train is faulty.

WARNING

Keep your weapon pointed toward the target.

Figure 3-17. Firing the launcher.

b. **Combat Environment**. If a misfire occurs in combat—

(1) Release the forward safety.

(2) Remove your right hand from the firing mechanism and cock the weapon again.

(3) Try to fire again. If the launcher still does not fire, maintain the same firing position and return the cocking lever to the SAFE (uncocked) position.

(4) Move the launcher from your shoulder, keeping the launcher pointed toward the enemy. Reinsert the transport safety pin.

(5) In combat, break off the sights to identify the misfired launcher. In training, however, you would not want to damage the field handling trainer (FHT), so identify the misfired launcher simply by leaving the sights up.

(6) Place the launcher on the ground, pointed toward the enemy, and use another launcher. As soon as you can, dispose of the misfired launcher IAW unit SOP.

c. **Training Environment**. If a misfire occurs on a live-fire training range, the soldier responds as follows (the trainer later disposes of the launcher IAW local SOP):

(1) Shouts "Misfire" as soon as the launcher fails to fire while maintaining the original sight picture.

(2) Releases the forward safety.

(3) Recocks the launcher: Immediately removes his right hand from the firing mechanism and pushes the cocking lever forward with the heel of his right hand until the lever locks with a loud clicking noise.

NOTE: Because performing immediate action takes so little time, you need not recheck the backblast area.

(4) Press the forward safety all the way down and try to fire again. If the launcher still fails to fire, shout misfire, release the forward safety, and move the cocking lever to the SAFE (uncocked) position. Move the launcher from your shoulder, keeping the weapon pointed toward the target and cradled in your left arm.

(5) Reinsert the transport safety pin, wait two minutes, then carefully lay the launcher on the ground, muzzle toward the target.

NOTE: Notify the local ammunition supply and issue point of any unusual occurrence, regardless of whether the weapon fires or not. Examples include excessive overpressure, recoil, or heat on your face after you have fired the weapon (caused by the propellant burning after the round leaves the muzzle).

3-9. RESTORATION TO CARRYING CONFIGURATION

If the launcher is prepared to fire, but then is not fired, it must be taken out of operation as follows:

a. Release the forward safety.

b. Push forward and to the left on the cocking lever, and let it spring back into the SAFE (uncocked) position.

c. Move the launcher from your shoulder, ensuring the muzzle is pointed in the direction of fire.

d. With the launcher cradled in your left arm, replace the transport safety pin until it is fully seated in the retainer hole.

e. To avoid breaking off the rear sight, remember to reset the range indicator to the 200-meter setting before closing the rear sight cover.

f. Lay down the sights and close their covers.

g. Snap the shoulder stop into the closed position.

h. Sling the launcher over your right shoulder and move to another location.

CHAPTER 4
MARKSMANSHIP FUNDAMENTALS

Many factors contribute to light antiarmor weapon marksmanship. Soldiers who combine these factors well, and continue to practice doing so, can retain their skills. The factors are grouped into four basic areas known as marksmanship fundamentals: steady hold, aiming procedures, breath control, and trigger manipulation. Instructions are given for right-handed firers, but the M72-series LAW can be fired from either shoulder by simply reversing the instructions.

4-1. STEADY HOLD

Maintaining a steady hold involves holding the launcher as steady as possible while sighting and firing. To maintain the proper sight picture and sight alignment until he fires, the firer must hold the launcher in a tight, comfortable position so that it becomes a natural extension of his body (Figure 4-1). With both weapons, keep your elbows close to your body to help balance the weapon and prevent you from jerking or flinching when you fire. In the case of the AT4, this reduces recoil.

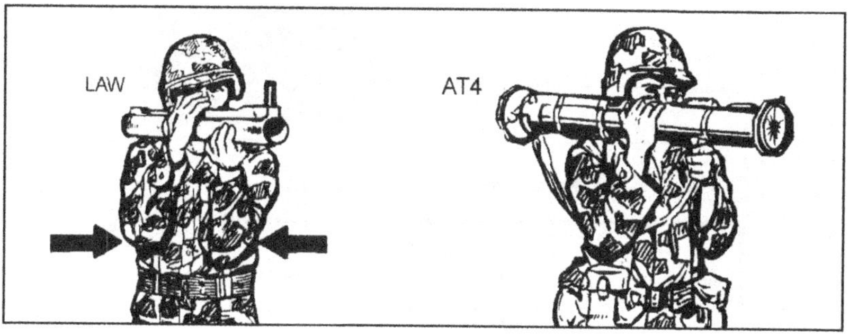

Figure 4-1. Steady hold position.

a. **M72-Series LAW**. Place your left hand, palm facing upward, under the launcher near the muzzle and grasp the launcher. Firmly pull the rear cover into your right shoulder pocket.

b. **M136 AT4**. With your left hand, grasp the carrying sling where it attaches to the launcher near the muzzle. With your right hand on the trigger mechanism, pull the shoulder stop into your right shoulder pocket.

4-2. AIMING PROCEDURES

Aiming procedures include placing the eye correctly, obtaining a sight picture, and aligning the sight. Combining these procedures is critical to correctly aiming light antiarmor weapons.

a. **Eye Placement**. Before sighting the weapon, estimate the range (Chapter 6 discusses range estimation). For the M72-series LAW, place your firing eye as close to the rear sight as is comfortable. However, for the M136 AT4, place your firing eye between 2 1/2 to 3 inches (no nearer than 2 1/2 inches) from the rear sight. This distance is necessary to prevent possible injury from the weapon's recoil and for correct sight alignment with the AT4 (Figure 4-2).

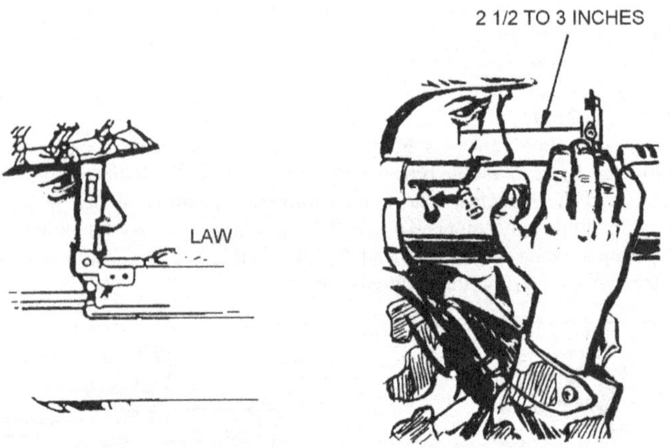

Figure 4-2. Eye placement.

> **WARNING**
> **When firing the M136 AT4, do not place your eye within 2 1/2 inches of the rear sight. The AT4s recoil could cause the rear sight to injure your firing eye.**

b. **Sight Alignment**. Align the sights correctly with the target. To do this for the M72-series LAW, position the rear sight so that your eye is near and in line with the peephole in the rear sight. Look through the peephole at the front sight reticle and place the range line that corresponds to the target's range on the target (Figure 4-3). To do this for the AT4, position the rear sight so that the white semicircle of the front sight is a hazy line around the bottom half of the rear sight opening. Position the front sight posts on the target (Figure 4-4). Align the sight by moving your head forward or backward.

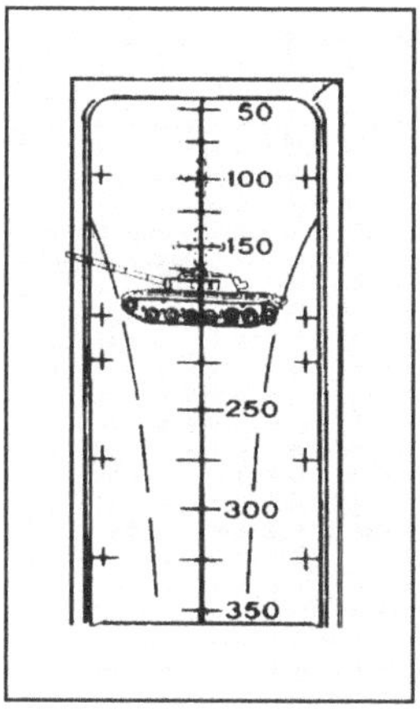

Figure 4-3. Sight alignment for the M72-series LAW.

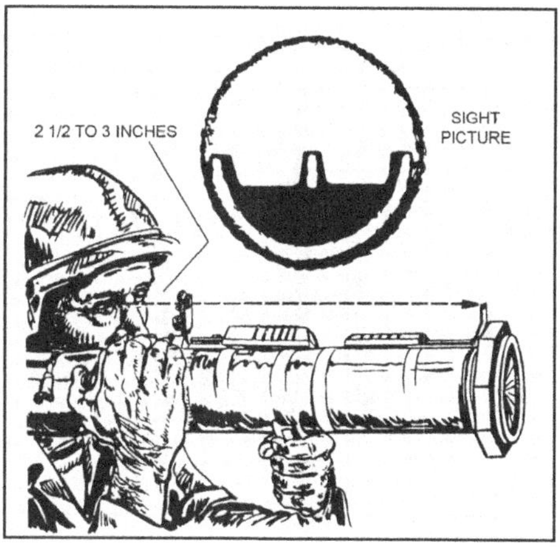

Figure 4-4. Sight alignment for the M136 AT4.

c. **Sight Picture**. Position the front sight on the target.

(1) ***M72-Series LAW***. *Stationary targets* include those moving directly toward or away from the firer. Place the correct vertical range line in the center of the target (Figure 4-5). *Slow-moving targets* include those with an estimated speed of 5 mph or slower, or those moving in an oblique direction. Place either the left or right lead cross mark on the vehicle's center of mass (Figure 4-6). *Fast-moving targets* include those with an estimated speed of more than 5 mph. Place either the left or right lead cross mark on the leading edge of the vehicle (Figure 4-7).

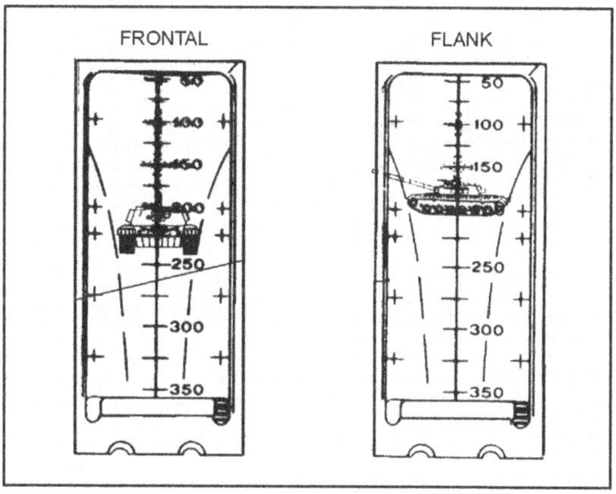

Figure 4-5. Sight picture, stationary targets, M72-series LAW.

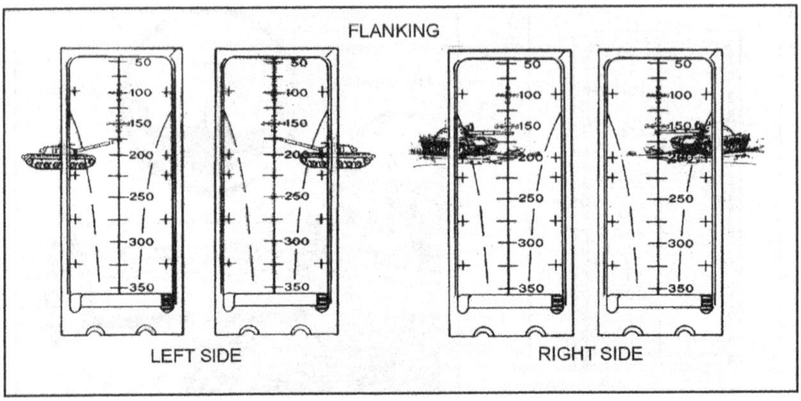

Figure 4-6. Sight picture, slow-moving targets, M72-series LAW.

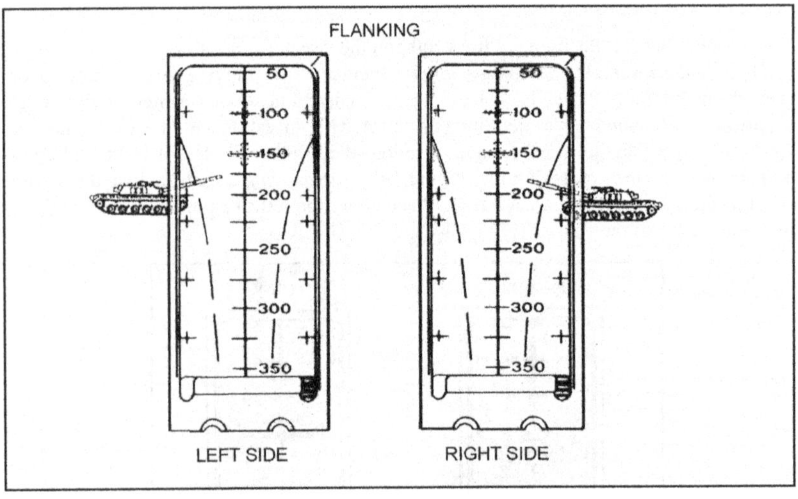

Figure 4-7. Sight picture, fast-moving targets, M72-series LAW.

(2) ***M136 AT4.*** *Stationary targets* include those moving directly toward or away from the firer. Adjust the rear sight for the correct range and place the center sight post in the center of the target (Figure 4-8). *Slow-moving vehicles* are those with an estimated speed of 10 mph or less, or those moving in an oblique direction. Place the center sight post on the front or leading edge of the vehicle (Figure 4-9). *Fast-moving vehicles* are those estimated to be moving faster than 10 mph. Place either the left or right lead post on the

center of the target. For example, if the target is moving from left to right, place the left lead post on the target's center of mass, and vice versa (Figure 4-10).

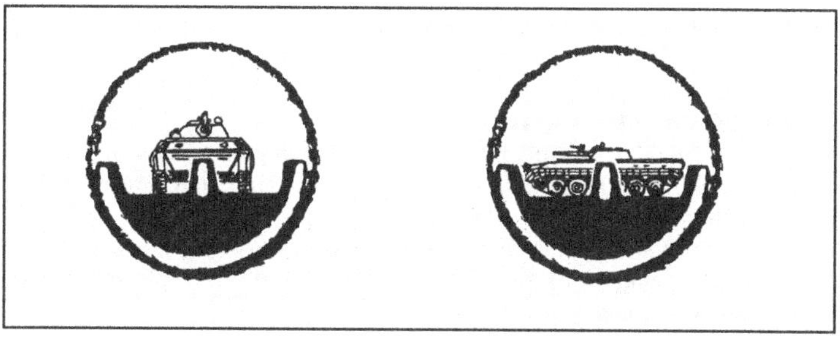

Figure 4-8. Sight picture, stationary targets, M136 AT4.

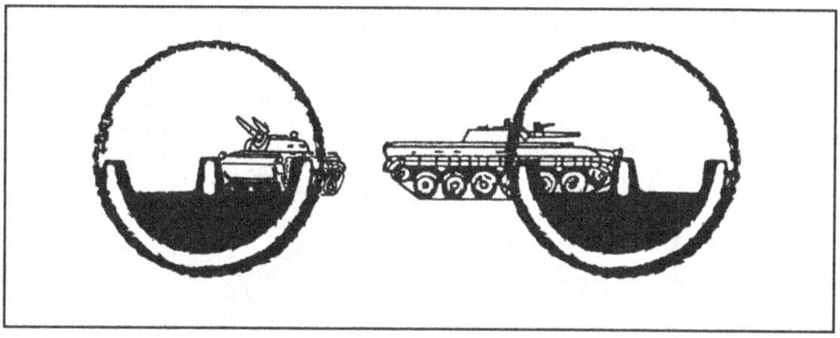

Figure 4-9. Sight picture, slow-moving targets, M136 AT4.

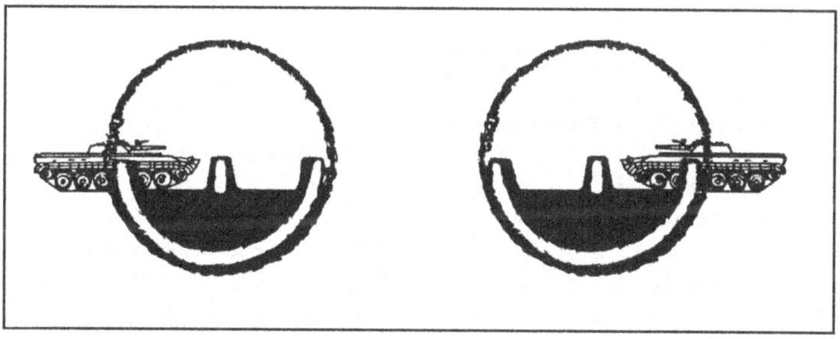

Figure 4-10. Sight picture, fast-moving targets, M136 AT4.

4-3. BREATH CONTROL

Breath control is as important when firing a light antiarmor weapon as it is when firing an individual weapon. Breathing while firing can cause a miss. To control breathing, the firer breathes deeply a couple of times, takes one last deep breath, exhales partly, holds his breath, sights, and fires.

4-4. TRIGGER MANIPULATION

Light antiarmor weapons have different types of triggers (Figure 4-11).

a. **LAW**. To fire the LAW, the firer must apply firm and steady downward pressure to the trigger with the fingers of his firing hand.

b. **AT4**. To fire the AT4, the firer must apply firm and steady forward pressure to the trigger with the thumb of his firing hand. Soldiers can practice trigger manipulation and control techniques on an expended launcher or FHT.

Figure 4-11. Trigger manipulation.

4-5. INTEGRATED ACT OF SHOOTING

Correct sight alignment is critical. Sight alignment errors increase as the range to the target increases. Therefore, maintaining the correct relationship between the rear and front sights is as important as placing the aiming point. The steps for doing this should become automatic. No matter how quickly they are done, these steps are always distinct, because the human eye can only focus at one distance and on one point at a time. The firer focuses on the front sight to obtain correct sight alignment, then places the aiming point to complete the sight picture. He shifts or adjusts the position of the launcher as necessary. The whole time he is pressing the trigger, he maintains the sight picture.

CHAPTER 5

FIRING POSITIONS

This chapter explains the basic firing positions used with light antiarmor weapons. Instructions for each are given for right-handed firers, but the M72-series LAW can be fired from either shoulder by simply reversing the instructions. Though each weapon can be fired from all four of the basic firing positions, individual physique determines exact body and hand positions. Firing from a supported position naturally increases accuracy, which improves the odds for a first-round hit or kill. Basic safety considerations are the same for all light antiarmor weapons, but additional considerations for each firing position are provided here.

5-1. STANDING POSITION

Two standing positions are used: a basic standing position and one modified for the infantry fighting position.

a. **Basic Standing Position**. Raise the launcher slightly higher than shoulder level. Execute a left face, rotate your shoulder under the launcher, and spread your feet a comfortable distance apart. Move your left foot 15 to 24 inches forward, keeping your hips level and your weight balanced on both feet. To obtain a firm, stable position, tuck both elbows tightly into your body. To track a moving target, turn your body at the waist—not with your legs. This enables you to track the target smoothly. Unless you are behind a protective barrier such as a wall, the standing position exposes you more than any other position to enemy observation and possible suppression. Differences between weapons with respect to the standing position are as follows (Figure 5-1, page 5-2):

WARNING

Always keep the launcher pointed in the direction of fire.

(1) ***M72-Series LAW***. Place your nonfiring hand about 4 inches from the front of the muzzle, with your firing hand on the rear cover. After placing the weapon on your shoulder, release the rear cover and place your firing hand on the trigger. Cup the launcher in the palm of your nonfiring hand. Position your firing eye as close to the rear sight as is comfortable.

(2) ***M136 AT4***. Grasp the sling near the launcher with your left hand and the shoulder stop with your right hand. Raise the launcher above shoulder level. After placing the launcher on your shoulder, release the shoulder stop and place your right hand on the trigger. Place your firing eye 2 1/2 to 3 inches from the rear sight.

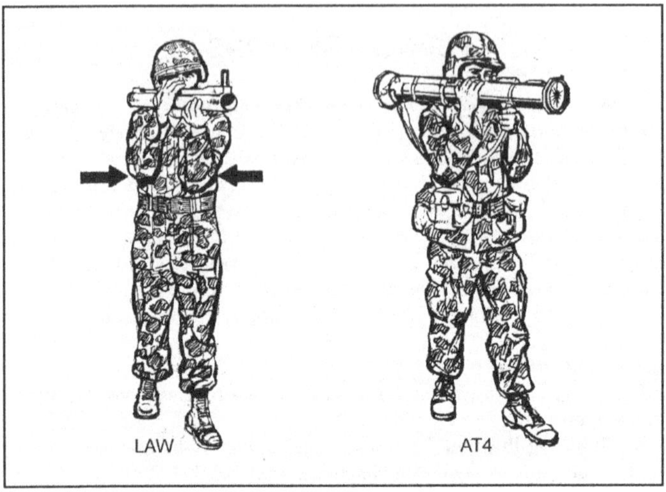

Figure 5-1. Basic standing position.

b. **Modified Standing Position**. Use this position when you occupy an infantry fighting position. Assume the basic standing position, but instead of stepping forward, lean against the back wall of the fighting position. Ensure that the venturi or rear of the weapon extends beyond the rear of the fighting position. Figure 5-2 shows the modified standing position for the M72-series LAW; Figure 5-3 shows the modified standing position for the M136 AT4. Ensure that NONE of the following are in your backblast area:

- Other soldiers.
- Other fighting positions.
- Equipment.
- Any part of your own fighting position.
- Obstructions within 5 meters.

NOTE: Leaders must ensure that light antiarmor weapons are positioned so that the backblast misses other fighting positions.

DANGER
NEVER FIRE FROM WITHIN A COMPLETELY ENCLOSED, UNVENTILATED BUNKER OR FIGHTING POSITION.

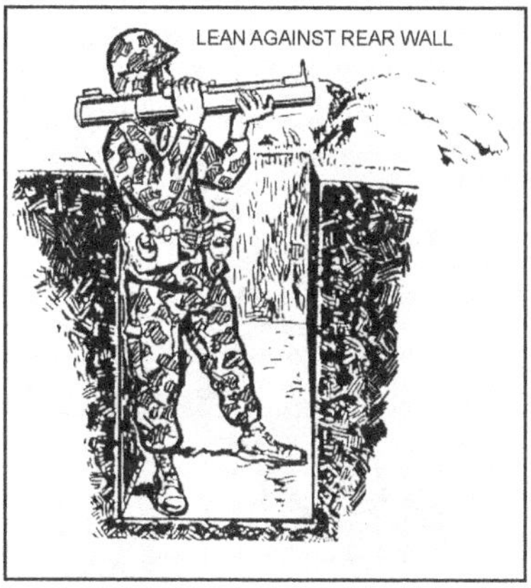

Figure 5-2. Modified standing, position, M72-series LAW.

Figure 5-3. Modified standing position, M136 AT4.

5-2. KNEELING POSITION

The basic kneeling position is the best position for tracking moving targets. The modified kneeling position is best for engaging stationary targets, since it is a supported position. However, either can be used for stationary or moving targets.

a. **Basic Kneeling Position**. Kneel from the basic standing position onto your right knee, keeping your left thigh parallel to the ground. Rotate your lower right leg 90 degrees to the left. (This removes your right foot from exposure to the backblast.) Keep your right thigh and back straight and perpendicular to the ground. Point your left foot in the direction of fire and tuck your elbows in to your sides. Though this is not a supported position, it should be a firm, stable one. Figure 5-4 shows the basic kneeling positions for the LAW and AT4.

b. **Modified Kneeling Position**. From the basic kneeling position, sit back on your right heel. Place the back of your upper left arm on your left knee, making sure you do not have bone-to-bone contact between your left elbow and left knee. Keep your right elbow tucked in close to your right side. Use any protective barriers available. Figure 5-5 shows the modified kneeling positions for the LAW and AT4.

Figure 5-4. Basic kneeling positions.

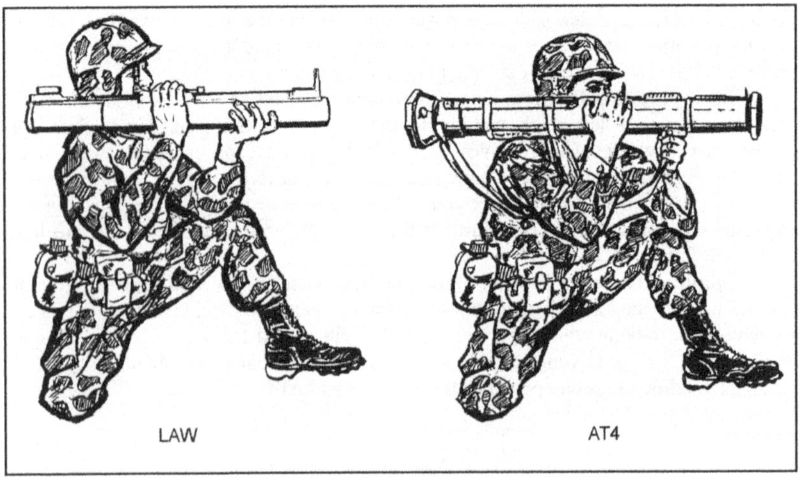

Figure 5-5. Modified kneeling positions.

5-3. SITTING POSITION

The sitting position is the most stable firing position. In this position, the arms are placed on the legs for support. Depending on his physique, the firer can use either of two versions of the sitting position. Either is suitable for engaging stationary targets.

a. **Basic Sitting Position**. Sit on your buttocks while facing the target, and spread your feet a comfortable distance apart. Lean forward and place the backs of your upper arms on your knees, avoiding bone-to-bone contact. Figure 5-6 shows the basic sitting positions for the LAW and AT4.

b. **Modified Sitting Position**. From the basic sitting position, cross your ankles for added support. Raise or lower your knees to adjust for elevation on the target. Figure 5-7 shows the modified sitting positions for the LAW and AT4.

Figure 5-6. Basic sitting positions.

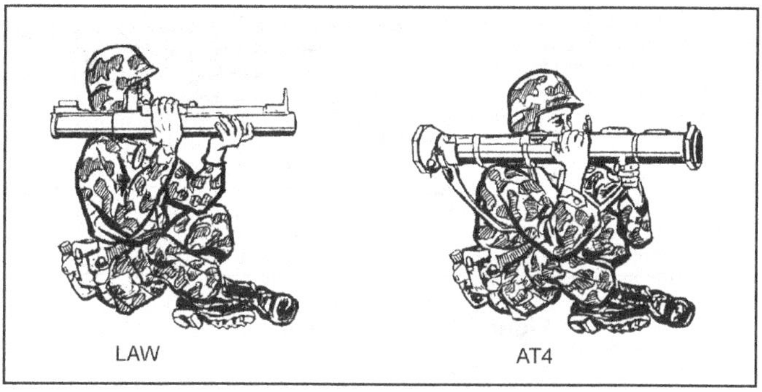

Figure 5-7. Modified sitting positions.

5-4. PRONE POSITION

The prone position is the most dangerous position due to its proximity to the ground. Ideally, the ground should slope downward from the rear of the launcher. This slope reduces the effects of the backblast.

a. Lie on your stomach with your body at a 90-degree angle to direction of fire, and with your body and legs to the left of the direction of fire.

b. *Ensure that neither the body nor the legs are in the backblast area.*

c. Unlike other firing positions, this one prevents you from placing the launcher on your right shoulder. Instead, you must hold the launcher in place against your upper right arm. For stability, apply extra pressure on the firing mechanism with your right hand. The prone position is the *least* stable of all firing positions. You must practice it often to become confident using it. Figure 5-8 shows the prone positions for the LAW and AT4.

DANGER

FAILURE TO MAINTAIN A 90-DEGREE ANGLE FROM THE DIRECTION OF FIRE COULD CAUSE INJURY OR DEATH TO THE FIRER.

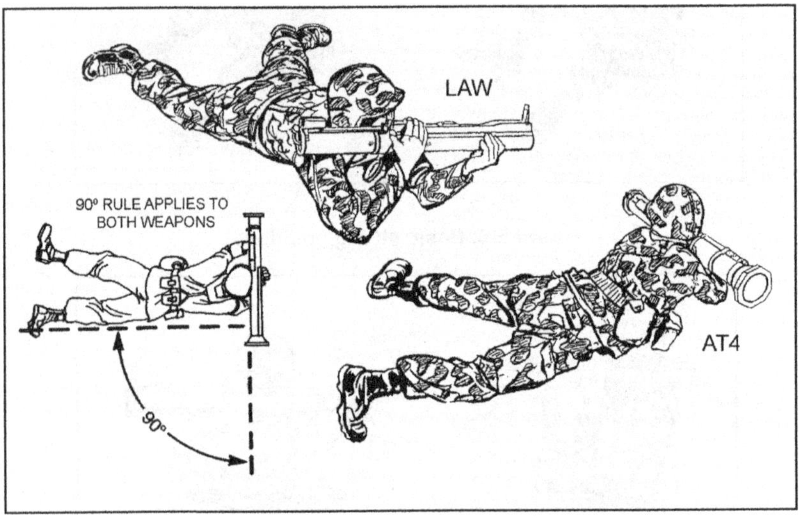

Figure 5-8. Prone positions.

CHAPTER 6
COMBAT TECHNIQUES

This chapter discusses employment techniques for light antiarmor weapons, all of which require at least basic gunnery skills. Techniques that require advanced skills are identified as such.

6-1. RANGE ESTIMATION

A firer who can accurately estimate the range to the target has a better chance of hitting it, regardless of the weapon used. Common methods of estimating range are listed below from the most to the least accurate. The tactical situation determines the method to be used:

a. Using range finders.

b. Measuring the distance on a map after correctly plotting your own position.

c. Pacing. Remember your individual pace count.

d. Using pair and sequence methods of target engagement. This method should be used only when in contact with the enemy.

e. Estimating range visually. This is the least accurate method of estimating range and therefore the least desirable. However, in an offensive operation or hasty defense, it may be the only method available to the light antiarmor firer. Thus, soldiers must continually train to improve their skill at visual estimation (STP 21-1-SMCT). Leaders should identify, coordinate, and record ranges to possible armored vehicle engagement locations on squad and platoon sector sketches.

6-2. SPEED ESTIMATION

Of the weapons discussed in this manual, the M136 AT4 is the best for engaging moving armored vehicles. One of its advantages over the LAW is the speed of its round, which travels faster and farther than the LAW round. However, the firer is the key in any engagement, especially a moving target engagement. Once soldiers learn to estimate speeds at known ranges, they should rehearse until they achieve a high hit-to-kill ratio. As their abilities improve, the leaders vary the ranges, speeds, and types of armored vehicles (Figure 6-1, page 6-2). Trainers and soldiers develop other methods through practice and are limited only by their imaginations. (Chapter 4 discusses obtaining a sight picture in detail.) Estimate how far the vehicle travels in 1 second:

a. Start when the front end of the vehicle passes the object.

b. Count, "One thousand and one" (takes about one second).

c. If *more* than half of the vehicle passes the object, estimate it as a *fast-moving* vehicle (10 mph or faster). If *less* than half of the vehicle passes the object, estimate it as a *slow-moving vehicle* (less than 10 mph).

Figure 6-1. Speed estimation.

6-3. ARMORED VEHICLE WEAKNESSES

Armored vehicles usually have their heaviest armor in front, because they are designed mainly for offensive operations against other armored vehicles (Figure 6-2). All vehicles are vulnerable to repeated hits on their flanks and rear, though the flank offers the largest possible target. Firers should always aim center of mass to increase the probability of a hit. The older the vehicle model, the less protection it has against antiarmor weapons. Consequently, newer versions may use bolt-on (applique) armor to improve their survivability. Some vehicles are equipped with reactive armor, which consists of metal plates and plastic explosives. Reactive armor usually covers the forward-facing portions and sides of the vehicle and can defeat shaped-charge weapons such as the LAW and AT4. When reactive armor detonates, it disperses metal fragments to 200 meters. The M72-series LAW and the M136 AT4 cause only a small entry hole in an armored vehicle target, though some fragmentation or spall may occur.

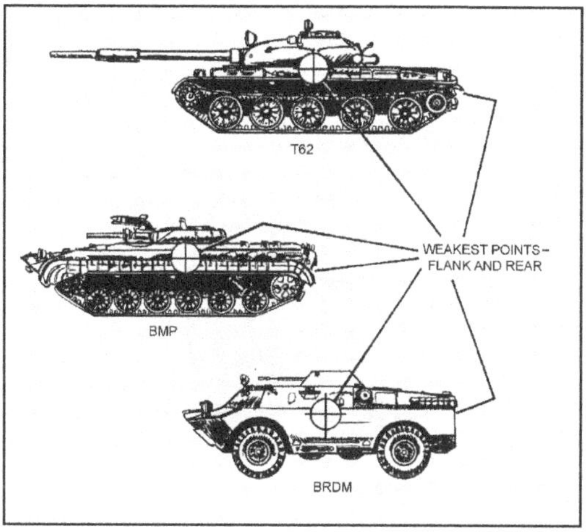

Figure 6-2. Armored vehicle weak points.

a. Natural or man-made obstacles can be used to force the armored vehicle to slow, stop, or change direction. This pause enables the firer to achieve a first-round hit. If he does not achieve a catastrophic kill on the first round, he or another firer must be ready to engage the target vehicle immediately with another round.

b. An armored vehicle without close protection (dismounted infantry) in woods, MOUT, or other restrictive terrain is vulnerable to close attack. This type of attack is most likely to originate from well-armed infantry-type teams organized into armor-killer teams. (Noninfantry units may also be required to perform this mission.) Skilled firers from these teams should engage the suspension or engine compartment of vehicles that have applique or reactive armor. When an armored vehicle is buttoned up—all hatches are closed and personnel are inside the vehicle—the crew cannot see well enough to protect itself from close attacks or attacks from the flanks or rear. The personnel inside cannot see anything within 10 meters of the vehicle, and they cannot shoot at anything (using their main guns) within 20 meters. The white area in Figure 6-3, page 6-4 shows the most favorable direction of attack when the turret is facing to the front; the gray area shows the vehicle's principal direction of fire and observation when the turret is facing to the front.

c. Armored vehicle kills are classified according to the level of damage achieved (Table 6-1).

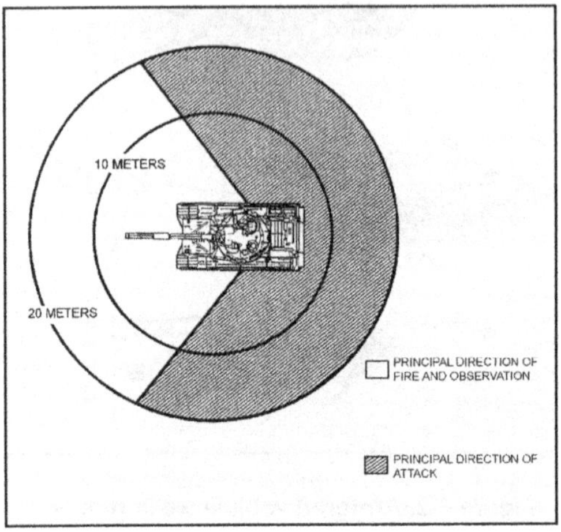

Figure 6-3. Limited visibility of armored vehicles.

TYPE OF KILL	PART OF VEHICLE DAMAGED OR DESTROYED	CAPABILITY AFTER KILL
Mobility Kill	Suspension (track, wheels, or road wheels) or power train (engine or transmission) has been damaged.	Vehicle cannot move, but it can still return fire.
Firepower Kill	Main armament has been disabled.	Vehicle can still move, so it can get away.
Catastrophic Kill	Ammunition or fuel storage section has been hit by more than one round.	Vehicle completely destroyed.

Table 6-1. Armored vehicle kills.

6-4. METHODS OF ENGAGEMENT

The four engagement methods include single, sequence, pair, and volley firing. The leader evaluates the situation on the ground to determine which of these methods to use. Regardless of whether they are used singly or in combination, communications are needed as well. The methods of engagement are rehearsed IAW unit SOP.

a. **Single Firing**. A single soldier with one light antiarmor weapon may engage an armored vehicle, but this is not the preferred method of engagement. Several light antiarmor weapons are required to kill an armored vehicle. A single firer firing one round must hit a vital part of the target to damage it at all (Figure 6-4).

(1) ***Range Not Known***. A single firer should engage only targets within 200 meters when he does not know the actual range. The probability that he hits a target beyond 200 meters with a single round is small.

(2) ***Range Known***. A single firer can engage targets out to 225 meters with the LAW or 300 meters with the AT4 when he knows the actual range. The firer should do this only when he has a flank or rear shot, or when he has no other engagement option.

Figure 6-4. Single firing.

b. **Sequence Firing**. A single firer, equipped with two or more light antiarmor weapons prepared for firing, engages the target. After engaging with the first round and observing the impact, the firer adjusts his point of aim, engages with another round, and so on until he destroys the target or runs out of rounds (Figure 6-5).

Figure 6-5. Sequence firing.

c. **Pair Firing**. Two or more firers, equipped with two or more light antiarmor weapons prepared for firing, engage a single target. Before firing, the first firer informs the others of the estimated speed and distance to the target. If the impact of his round proves his estimate to be correct, the other firers engage the target until it is destroyed. If the impact of the round proves his estimate to be incorrect, the second firer informs the others of his own estimate, then he engages the target. This continues until the target is destroyed or all rounds are expended (Figure 6-6).

Figure 6-6. Pair firing.

d. **Volley Firing**. Two or more firers can engage a single target when the range is known. These firers engage the target at one time on a prearranged signal such as a command, whistle, booby trap, mine, or TRP. For the best method of engagement, use the light antiarmor weapon, because it places the most possible rounds on one target at one time, increasing the possibility of a kill (Figure 6-7).

Figure 6-7. Volley firing.

e. **Communications**. Leaders control all unit fire and communicate this information to the entire unit IAW unit SOP. Light antiarmor weapons firers must know—

- Designated firers.
- Target priority.
- Method of engagement.
- Range and lead to target (if known).
- Command or signal to fire.
- Command or signal to cease fire.

6-5. ENGAGEMENT OF FIELD FORTIFICATIONS AND BUILDINGS

Light antiarmor weapons have little effect against field fortifications and buildings. Soldiers should not expect to severely damage targets with these weapons. However, if the alternatives shown in Table 6-2, page 6-8, are used, soldiers may be able to gain a temporary advantage.

6-6. ENGAGEMENT OF OTHER VEHICLES

The M72-series LAW proves more effective against light vehicles; the M136 AT4 proves more effective against armored vehicles. Nonarmored vehicles, such as trucks, cars, and boats, are considered *soft targets*. Firing along their length offers the greatest chance of a kill, because this type of shot is most likely to hit their engine block or fuel tank.

AIM POINT	EFFECT WHEN WEAPON IS FIRED AT AIM POINT	RECOMMENDED FIRING TECHNIQUE
BUNKER OR FIGHTING POSITION		
Firing Port or Aperture	Rounds fired into firing ports or apertures are wasted: rounds detonate inside rear of position, causing little structural damage to the position or to the equipment or personnel within, unless they are hit directly. The AT4 produces less effect than the LAW.	Coordinate fire: Fire *light antiarmor weapons* at a point 6 to 12 inches from the front edge of the firing ports in the berm. Fire *small arms* at the bunker or position to prevent personnel within from returning light antiarmor fire.
Berm	Firing at the berm causes the round to detonate outside the fighting position or inside the berm, creating only a small hole in the berm, dust, or minor structural damage to the position, but no damage to personnel or equipment unless they are hit directly. The AT4 produces less effect than the LAW.	
Window	The round may travel completely through the structure before detonating; if not, it creates dust and causes minor structural damage to the rear wall, but little damage to personnel or equipment, unless they are hit directly. The AT4 produces *less* effect than the LAW.	Fire 6 to 12 inches from the sides or bottom of a window. Light antiarmor rounds explode on contact with brick or concrete, creating an opening whose size is determined by the type of round used.
Wall	The round detonates on contact, creating dust and causing a small hole and minor structural damage, but little damage to personnel or equipment, unless they are hit directly. The LAW may be used to create a loophole, which is a hole large enough to throw hand grenades through. The AT4 produces *less* effect than the LAW.	
Corner	Corners are reinforced and thus harder to penetrate than other parts of a wall. Any light antiarmor round will detonate sooner on a corner than on a less dense surface. Detonation should occur in the targeted room, creating dust and causing *overpressure*, which can temporarily incapacitate personnel inside the structure near the point of detonation. The AT4 causes *more* overpressure than the LAW.	

Table 6-2. Effects of light antiarmor weapons on field fortifications or bunkers.

6-7. LIMITED VISIBILITY ENGAGEMENTS

Limited visibility engagements can be conducted using various night vision devices or with artificial illumination. However, when NVDs or artificial illumination is used, limited visibility can reduce the maximum effective range for light antiarmor weapons by at least one-third. To avoid fratricide, leaders must ensure all designated light antiarmor weapon firers are trained to use their weapons in limited visibility.

a. **Night Vision Device**. Before a NVD can be used with the M136/AT4, it must be removed from its designated weapon (M249 machine gun or automatic weapon, or M60 machine gun) and the M136/AT4 mounting bracket must be attached. Appendix F describes the various NVDs that can be used, it also gives information for mounting, boresighting, and zeroing procedures for each NVD.

b. **Artificial Illumination**. If artificial illumination is used during a limited visibility engagement, it should be placed above and slightly beyond the target. However, the ability to identify and engage targets is even less with artificial illumination than with NVDs.

6-8. ENGAGEMENT IN NBC CONDITIONS

Wearing a protective mask limits the firer's ability to sight the weapon. Wearing NBC gloves limits his ability to manipulate the firing mechanism.

a. **Sighting the Weapon**. Sighting while wearing the protective mask may require rotating the weapon slightly counterclockwise. The mask also makes determining the location, identity, and engageability of targets more difficult.

b. **Firing the Weapon**. Practice manipulating the firing mechanism while wearing NBC gloves.

NOTE: When live firing either a light antiarmor weapon or its subcaliber trainer, aim within range firing limits.

DANGER

THE M136 AT4 IS NOT RATED SAFE. IN TRAINING, NEVER FIRE IT FROM AN ENCLOSURE; IN COMBAT, DO SO ONLY WHEN NO OTHER POSITION IS AVAILABLE. THE OVERPRESSURE AND BLAST COMBINED CREATE CONDITIONS THAT CAN KILL YOU. IF THEY DON'T KILL YOU, THEY CAN INJURE YOU SERIOUSLY OR DEAFEN YOU TEMPORARILY OR PERMANENTLY.

6-9. ENGAGEMENT FROM AN ENCLOSURE

Firing from an enclosure creates unique hazards. As such, before positioning soldiers in enclosures (in combat only), leaders must consider several factors that affect safety. Only in combat, *when no other tactical option exists*, should the M136 AT4 be fired from an enclosure. If it must be employed this way, the enclosure must meet the following minimum requirements. The M72-series LAW has been rated as safe for use from an enclosure but, again, only when the enclosure meets the following minimum requirements:

a. **Construction**. The building must be sturdily constructed to reduce the structural damage that would occur in a weakly constructed enclosure such as one made of wood or stucco.

b. **Size of Enclosure**. Minimum measurements for the enclosure are as follows:

- AT4—minimum room size 17 x 24 feet.
- LAW—minimum room size 12 x 15 feet.
- Both—minimum ceiling height 8 feet.

c. **Ventilation to the Rear and Sides**. To allow for the backblast, at least 20 square feet of ventilation—such as a standard 3-foot by 7-foot doorway—must be provided directly behind the firer. More doors and windows are removed beside and behind the position to increase ventilation and reduce overpressure, noise, and blast effects. Without sufficient ventilation, the blast would weaken or collapse the walls. On the front wall, windows and doors are reinforced rather than removed, because removing would draw attention to the position. Reinforcing the windows also helps protect the firer from enemy direct-fire weapons.

d. **Objects and Debris**. All objects and debris are removed from the rear of the weapon, because the backblast causes them to fly around the room injuring personnel.

e. **Muzzle Clearance**. Muzzle Clearance must be at least 6 inches.

f. **Weapon Clearance**. Properly positioning the weapons within the enclosure is vital to the safety and survival of all personnel in the enclosure. The weapons should be positioned so that no walls are within 5 meters to the rear or side of the weapon.

g. **Personnel Positions**. If any other soldiers must be present, then they must remain forward of the rear of the launcher and avoid standing in corners or near walls. If possible, they should construct reinforced positions that fit the previous criteria and that can protect them in case the building collapses.

WARNING

To avoid injuring the eardrums, soldiers must wear the approved brand of ear protection.

6-10. ENGAGEMENT BEYOND MAXIMUM EFFECTIVE RANGE (M136 AT4 ONLY)

A skilled M136 AT4 firer can engage targets beyond the weapon's maximum effective range of 300 meters, up to 550 meters. Beyond 550 meters, the firer must aim higher than center of mass and apply additional lead for moving targets. Commanders must realize that accuracy is reduced at these ranges. Also, firing at these ranges identifies the firing position to the enemy.

6-11. OFFENSIVE OPERATIONS

All elements, even those with other organic antiarmor weapons, use light antiarmor weapons. Light antiarmor weapons can influence the action in an attack, so units should routinely stock them beforehand. They are most useful against lightly armored vehicles. They can also be used against soft targets, such as bunkers, field fortifications, automobiles, and trucks, but their shaped-charge warheads have less effect on these than on armored targets. Unless personnel, ordnance, or flammable material on or inside them are hit, soft targets can normally continue to fight after being attacked by light antiarmor weapons. Due to their relatively short range, light antiarmor weapons should be placed throughout the attacking force. They support the maneuver by providing a base of fire, and they enable the assaulting force to engage in close antiarmor combat.

6-12. DEFENSIVE OPERATIONS

Whether or not other organic antiarmor weapons are available, light antiarmor weapons are an asset in the squad and platoon defensive plan.

a. Light antiarmor weapons are employed with interlocking fires to provide mutual support. Dispersion allows leaders the flexibility to place flank, rear, and oblique fires on targets. This procedure increases the survivability of the firers as well as the probability of achieving kills. Leaders must select positions that avoid fratricide from antiarmor backblast and short rounds (Appendix A).

b. The squad and platoon leader's sector sketch identifies TRPs and primary areas of possible engagement. The sector sketch also identifies possible avenues of approach for enemy armored vehicles. Each light antiarmor weapon shares a sector of fire with the primary small-arms weapon assigned to its position. However, because it is neither a primary nor crew-served weapon, the light antiarmor weapon does not require a separate range card. When assigning sectors of fire, squad and platoon leaders inform firers of all possible target areas, TRPs, and prearranged signals (Figure 6-8). (FMs 7-7, 7-7J, and 7-8 provide more information about the squad and platoon sector sketch.) Leaders provide each designated position with two or more light antiarmor weapons.

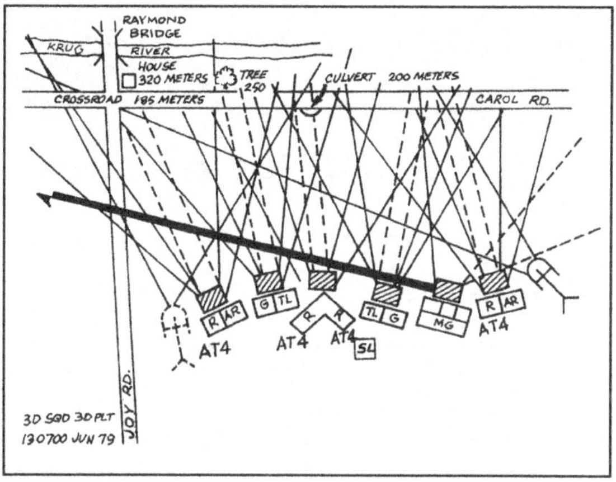

Figure 6-8. Squad sector sketch.

6-13. OTHER TACTICAL OPERATIONS

The weight of the light antiarmor weapon suits it well for combat patrols and rear area operations. (FM 7-8 provides more information about these subjects.)

a. **Combat Patrols**. Light antiarmor weapons are used on combat patrols to destroy enemy equipment, installations, and key points, and to harass enemy forces. The two types of combat patrols are—

(1) ***Ambushes (Antiarmor)***. The commander can employ armor-killer teams either during limited visibility or when cover, concealment, and withdrawal routes are available. The key to ambushing armored or other vehicles is to choose terrain that restricts their maneuverability and fields of fire, but that allows friendly forces to engage the vehicles from the flank and rear. Soldiers can emplace antipersonnel mines before the ambush so that dismounting enemy soldiers deploy into them. Though light antiarmor weapons can be used independently, they are normally used in support of designated organic antiarmor weapons such as the Dragon. Volley firing light antiarmor weapons increases the probability of a quick kill (Figure 6-9). Security teams stop the enemy from escaping.

Figure 6-9. Antiarmor ambushes.

(2) ***Raids***. Using light antiarmor weapons on raids increases target options. Light antiarmor weapons can also be used to defend against armored vehicles.

b. **Rear Area Operations**. Rear area operations usually consist of either: defending unit trains and tactical operation centers, or patrolling rear areas.

(1) ***Defense of Tactical Operations Center and Unit Trains***. The mobility of modern vehicles makes the TOC and unit trains prime targets. Light antiarmor weapons are used to engage vehicles threatening the unit during the defense of the TOC. Soldiers who use light antiarmor weapons in this type of operation normally perform noninfantry-type roles. Unit leaders designate soldiers to carry light antiarmor weapons and ensure that these soldiers receive light antiarmor weapons training.

(2) ***Patrols of Rear Areas***. Rear area patrols are security patrols conducted by designated infantry or MP units. These patrols can react to any threat they encounter in the rear area. Rear area patrols use the light antiarmor weapon in a hasty point defense at a roadblock, intersection, or strongpoint.

CHAPTER 7

TRAIN-THE-TRAINER AND UNIT TRAINING PROGRAMS

During initial-entry training, soldiers receive only an orientation for each of the light antiarmor weapons. When they arrive in a unit, soldiers should train to Skill Level 1 proficiency on all assigned weapons and equipment. Each unit should have an ongoing unit training program to allow soldiers to work together to attain and sustain their skills. This program integrates individual and collective tasks. Operating a unit training program requires the unit to also maintain a train-the-trainer program. Both programs are planned and conducted IAW the commander's METL.

7-1. TRAINING STRATEGY

Training strategy integrates resources into a year-round program to train the individual and collective skills needed to perform the wartime mission. Thus, units are trained to fight and win on the battlefield. The training strategy includes institutional and unit training.

a. **Institutional Training.** TRADOC schools conduct institutional training. These courses train critical individual and leader skills required to succeed in combat. Students are provided the latest doctrine and taught current technical skills associated with the light antiarmor weapons available within the Army. Soldiers train at several institutional schools throughout their career (see the appropriate programs of instruction [POIs]) (Table 7-1).

ENLISTED	Initial-Entry Training
	Primary Leadership Development Course
	Basic Noncommissioned Officer Course
	Advanced Noncommissioned Officer Course
OFFICERS	Precommission Training
	Officer Basic Course
	Officer Advanced Course

Table 7-1. Institutional training.

b. **Unit Training.** Unit training consists of sustainment training in individual and collective skills. Commanders manage their resources (devices, simulators, simulations, ranges, and ammunition) carefully to best develop individual and leader skills. Through the use of exercises, (drills, STXs, and qualification tables), the unit develops skills and integrates individuals into a cohesive crew or squad. Table 7-2 provides a training guide for building a light antiarmor weapons sustainment training plan (see also Appendixes C and E).

7-2. UNIT TRAINING PROGRAM

Every unit commander, including those commanding CS and CSS units, should develop a program for light antiarmor training. Some units have more than one type of light antiarmor weapon that requires sustainment training. The commander considers preparation and training for both collective and individual tasks. Trainers analyze their collective tasks and the individual tasks that support them. Individual tasks must be integrated into collective training and rehearsals. The units should try to train light antiarmor weapons at the same time and on the same range (Appendix D). Otherwise, these weapons are trained in alternating quarters of the year.

a. **Sustainment Training Requirements**. Sustainment training is recommended for each soldier. Active Component units should conduct sustainment training quarterly and fire the appropriate qualification tables semiannually, depending on the STRAC TRC level. Reserve Component units should conduct sustainment training semiannually and fire the appropriate instructional or qualification tables annually. All units should fire advanced instructional tables semiannually. When possible, they should conduct live fire annually IAW the current STRAC manual. (Appendix E provides all the necessary firing tables.) Commanders should ensure that designated gunners complete unit sustainment training and live fire tactical light antiarmor weapons sometime during that quarter. *Training requirements* are divided into four tracks, A through D, based on ammunition allocations (Table 7-3). Table 7-4 provides an example *training frequency chart*. This example is based on the AT4. The same tasks apply to all light antiarmor weapons, but procedures and standards may vary. Tactical weapons are allocated IAW the current standards in training commission (STRAC) manual.

1	Designated soldiers should train as often as needed to maintain proficiency.
2	Performance evaluations should be administered to designated soldiers quarterly, if resources allow.
3	Designated soldiers should fire the qualification tables semiannually for verification on designated weapons.
4	Designated soldiers qualify semiannually on their designated weapons. Qualification consists of a hands-on performance evaluation and qualification firing tables.
5	Designated soldiers should be trained in advanced *instructional* techniques as often as the unit training program allows. This training helps to maintain their target engagement skills.
6	All soldiers not designated as light antiarmor weapons gunners should train on these weapons to the extent allowed by resources.

Table 7-2. Sustainment training plan for light antiarmor weapons.

TRACK	DETERMINES ANNUAL AMMUNITION ALLOCATION FOR:
A	Each Active Component designated firer.
B	Track deleted
C	Each Reserve Component designated firer.
D	Each Reserve Component training division instructor.

Table 7-3. Training requirements.

TRAINING OBJECTIVE	TASK	FREQUENCY		APPROXIMATE TIME
		AC	RC	
NA	Conduct an orientation safety briefing (as appropriate).	Q	A	10 min
1	Inspect an M136 AT4 for serviceability.	Q	A	30 min
2	Prepare an M136 AT4 for firing.	Q	A	45 min
3	Demonstrate correct M136 AT4 firing positions.	Q	A	45 min
4	Estimate range.	Q	A	60 min
5	Determine the correct M136 AT4 sight picture.	Q	A	30 min
6	Perform M136 AT4 misfire procedures.	Q	A	45 min
7	Return the M136 AT4 to the carrying position.	Q	A	45 min
8	Engage targets with an M136 AT4.	S	B	3 1/2 hr
KEY: **Q = quarterly (every 90 days)** **S = semiannually (twice a year)** **A = annually (once a year)** **B = biennially (every two years)**				

Table 7-4. Example training frequency chart.

b. **Gunnery Training Requirements**. All designated soldiers should be trained to standard IAW the combat techniques in Chapter 6 and the performance evaluations in Appendix C.

c. **Collective Training.** Individual tasks must be integrated into collective training and rehearsals. To accomplish this, commanders analyze the collective tasks from their unit METL and the individual tasks that support the collective tasks. The commander determines the collective tasks that must be practiced by analyzing the mission outlines in the appropriate mission training plan (MTP). The mission outlines describe the collective tasks required to execute each ARTEP mission. Leaders determine the individual tasks that support the collective tasks by referring to the mission task matrix in the appropriate MTP.

(1) Leaders use the training standards given in the appropriate manual for the individual tasks. Noninfantry units can use FMs 7-7, 7-7J, 7-8, 7-10, and 7-20 to identify infantry-type missions. Some of the tasks identified may include engaging armored vehicles. When planning training, the trainer incorporates the training devices and aids for light antiarmor weapons into the unit's training. These devices and aids add realism and allow trainer to properly evaluate the task being conducted. He must also identify deficiencies and correct them by retraining soldiers to use the selected weapons and their related training devices correctly.

(2) The MILES Viper is a force-on-force trainer for light antiarmor weapons. MILES trainers are *not* gunnery trainers and are not used for sustainment training. They are used in all force-on-force training. Table 7-5 provides an example MILES training program.

SEQUENCE	FREQUENCY (TIMES PER YEAR)	
	ACTIVE	RESERVE
Squad, platoon, or company FTX or STX	4	1
Battalion FTX	2	0
Battalion ARTEP	2	1
Total MILES training events per year	8	2

Table 7-5. MILES training program.

7-3. TRAIN-THE-TRAINER PROGRAM

The success of the unit's light antiarmor training program depends on trainers who are well-trained, competent, and prepared to train soldiers.

a. **Objectives**. The train-the-trainer program must teach trainers to do the following:

- Evaluate the unit's training weaknesses.
- Plan training.
- Set up, operate, and maintain training equipment.
- Conduct training.
- Coach firers in gunnery training.
- Evaluate firers' training weaknesses.
- Correct firers' training weaknesses.
- Instill confidence in the firers trained.
- Maintain training records.

b. **Mission-Essential Task List**. The commander must carefully examine his unit's wartime mission and develop a METL. He must then organize the unit's sustainment training program to support the METL. The light antiarmor sustainment training plan consists of quarterly, semiannual, and annual training events. Between these events, the unit provides additional light antiarmor weapons training. The sustainment plan should prepare the unit for the next round of weapons gunnery while also supporting the unit METL.

c. **Trainer Assessment**. Trainers within a unit are normally team, squad, and section leaders and platoon sergeants. Before becoming weapons trainers, they must be assessed carefully and any shortcomings corrected. The commander chooses a method of assessing the trainers that ensures that their abilities are accurately evaluated. Then, with the assistance of the platoon leaders and platoon sergeants, the commander performs this assessment.

d. **Command Benefits**. Regular and aggressive train-the-trainer programs provide valuable benefits that gradually spread through the unit. These benefits include a base of expertise, proficiency, and esprit de corps. Maintaining this base is easier than recreating it.

e. **Training Tasks**. Successful trainers know how to operate the training devices for the light antiarmor weapons assigned to their units. Based on the unit's METL, the commander designates the appropriate light antiarmor weapon for specified missions. Appendix C provides the tasks, conditions, and standards for each weapon discussed in this manual. The trainers must know the appropriate combat techniques for employing these weapons. They must also perform the following:

- Conduct an orientation safety briefing.
- Inspect the weapons for serviceability.
- Prepare the weapons for firing.
- Demonstrate the correct firing positions.
- Estimate range.
- Determine the correct sight picture.
- Perform the correct combat and training misfire procedures.
- Return the weapon to the carrying configuration.

APPENDIX A
SAFETY

This appendix provides safety procedures for both training and combat.

A-1. TRAINING SAFETY

Training safety for light antiarmor weapons applies to all types of ranges and training sites.

a. **Backblast**. Light antiarmor weapons have little or no recoil, because the propellant gasses escape to the rear of the weapon. This backblast can damage equipment or seriously injure personnel who are too close to the rear of the launcher. When operating temperatures fall below freezing (0NC or 32NF), all backblast areas and safety zones double.

(1) ***M72-series LAW***. This is an open-chambered weapon, so it has no recoil. The launcher's total backblast area extends 40 meters (44 yards) to the rear (Figure A-1) and is divided into two zones. During training, both should be marked off limits:

(a) Danger zone. All personnel, equipment, and flammable material must be clear of this area.

(b) Caution zone. The weapon's backblast may throw loose objects to the rear. Therefore, personnel must also stay clear of this area.

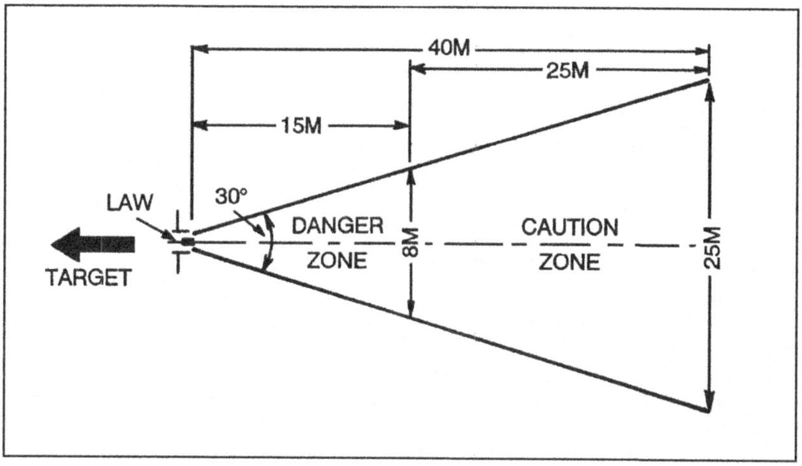

Figure A-1. M72-series LAW backblast area.

(2) ***M136 AT4***. Because the AT4 has a closed chamber, it has a slight recoil. When the weapon is fired, pressure builds up in the chamber and ruptures the baseplate. Once the baseplate ruptures, gasses exit from the back of the weapon, forming the backblast. The total backblast area extends 100 meters to the rear of the launcher in a 90-degree fan (Figure A-2, page A-2). During training, the *entire* backblast area should be marked off limits and kept clear of personnel and equipment. In combat, personnel may remain to the rear of the

weapon, but *only* if they are in protected positions. Backblast heat or flying debris could injure the firer and other soldiers anywhere in the backblast area. This weapon must not be fired from an enclosure or in front of a barrier (TM 9-1315-886-12). It may be fired from an enclosure in combat *only when* no other tactical option exists.

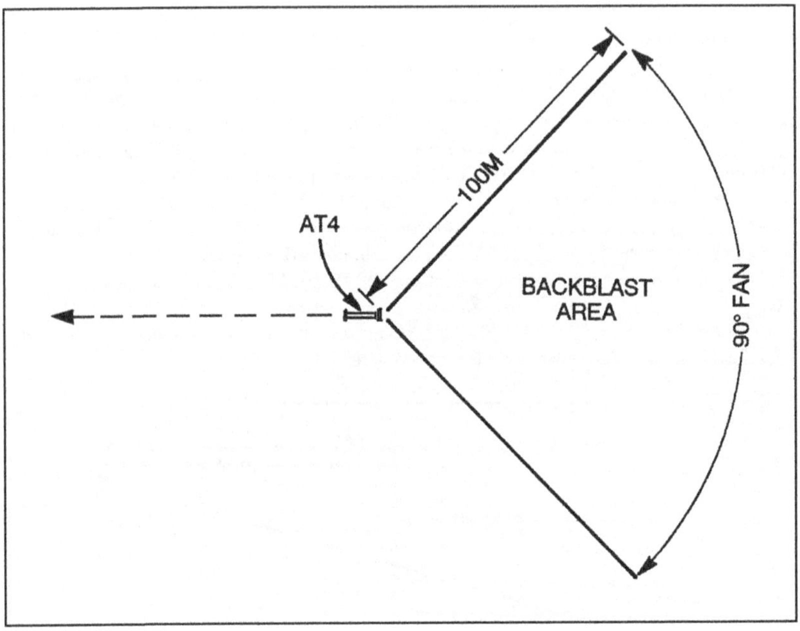

Figure A-2. M136 AT4 backblast area.

b. **Firing Limitations**. Trainers and soldiers must observe the backblast danger zones as well as other firing limitations. A light antiarmor weapon can be fired from the standard infantry fighting position. However, if this position is chosen, trainers/soldiers must consider other factors that increase the accuracy and reduce the danger to friendly soldiers: Rear of launcher extends beyond the back wall of the position, no overhead cover extends into the backblast area, and at least 1/2 meter (18 inches) clearance is around the weapon.

c. **Range Procedures for Light Antiarmor Weapons**. For all the light antiarmor weapons discussed in this manual, these procedures must be followed:

(1) The whole backblast area must be well marked and must be clear of all personnel, equipment, and flammable materials.

(2) No one enters the area behind the firing line or forward of the rear safety line without permission from the range OIC or safety NCO.

(3) Weapons are stored to prevent exposure to the sun or extreme temperatures.

d. **Additional Range Procedures for the M136 AT4**. The extreme overpressure and noise created when this weapon is fired require special precautions during training. Not only must firers and others present wear earplugs, but they may only use the approved brand—no other brand provides sufficient protection. To reduce these hazards, the range firing limitations shown in Table A-1 must be followed.

Within a 24-hour period, a soldier may only fire, observe fire, or act as safety NCO for the M136 AT4 one to three times, depending upon the firing position used	3 times (total) if the AT4 is fired from either the standing or kneeling position OR 2 times if the AT4 is fired from the sitting position OR 1 time (total) if the AT4 is fired from either the prone or the fighting position

Table A-1. M136 AT4 firing limitations during training.

DANGER

DURING TRAINING, DO NOT FIRE LIGHT ANTIARMOR WEAPONS FROM ANY ENCLOSURE OR FROM BEHIND ANY BARRIER.

A-2. COMBAT SAFETY

Combat safety rules and procedures include all those that apply to training plus the following:

a. **Fighting Position**. Both the M72-series LAW and the M136 AT4 can be fired from the standard infantry fighting position. However, to increase accuracy and reduce the danger to friendly soldiers, the area to the rear of the firing position must have no walls, large trees, or other obstructions within 5 meters (5 1/2 yards) to deflect the backblast onto the firer or into the position.

DANGER

ENSURE THE AREA TO THE REAR OF THE FIRING POSITION HAS NO WALLS, LARGE TREES, OR OTHER OBSTRUCTIONS WITIN 5 METERS (5 1/2 YARDS). OBSTRUCTIONS DEFLECT BACKBLAST ONTO THE FIRER OR INTO THE POSITION, INJURING OR KILLING THE FIRER AND ANY OTHER SOLDIER OCCUPYING THE FIRING POSITION.

(1) In the *individual infantry fighting position*, the soldier must lean against the rear wall and ensure that the venturi or the rear of the weapon protrudes past the rear of the position. The firer's elbows are not supported.

(2) In the *two-soldier infantry fighting position*, the nonfiring soldier must remain clear of the backblast area. These positions should be constructed and sited so that none are located in another's backblast danger zone.

(3) A *modified firing position* may be constructed to the side of the two-soldier fighting position. Firing from a modified position reduces the possibility of injury to the firer or another soldier in the fighting position, but it offers the firer some protection from enemy return fire.

b. **Temperature Limitations**. No light antiarmor weapon should be fired when its temperature exceeds operating limits. When a weapon temperature is below freezing (0 degrees Centigrade or 32 degrees Fahrenheit), overpressure increases and prevents some of the propellant from burning. Then, as the projectile leaves the launch tube, portions of the remaining propellant could be directed onto the firer's face or upper body. Consequently, in cold weather, the firer should wear a protective mask. On the other hand, when the weapon temperature is above 60 degrees Centigrade (140 degrees Fahrenheit), propellants and warhead compounds can destabilize, so, whether they are located at training sites or fighting positions, weapons should be stored in a shaded area in hot weather. Finally, weapons must not be fired when their temperatures exceed safe operating limits: -40 degrees to 60 degrees Centigrade (-40 degrees to +140 degrees Fahrenheit).

c. **Minimum Target Engagement Ranges**. In combat, no targets will be engaged within the 10-meter minimum arming range. During training, no targets will be engaged within 30 meters.

d. **Overhead Fire**. Light antiarmor weapons must not be fired over the heads of friendly soldiers, unless the soldiers have adequate protection against direct impact or other hazards.

APPENDIX B

TRAINING DEVICES AND AIDS

Training devices and training aids enable soldiers to learn as much as they can about a weapon before they try the real thing. This procedure saves money and time, and it prevents injuries. ***Training devices*** *both look and behave like the real weapon, but are cheaper and safer to practice with.* ***Training aids*** *include anything else used to help soldiers learn to use a weapon. Trainers should use their imaginations and invent or adapt other training aids from available resources.*

B-1. LAW SUBCALIBER TRAINING DEVICE

The only M72-series LAW training device designed for individual use is a subcaliber trainer (Figure B-1). This device, referred to as the M190 subcaliber launcher, is made by adding an M190 subcaliber conversion kit (NSN 1340-00-420-7999) (Figure B-2) to an expended M72-series LAW launcher. It is used to fire the M73 subcaliber (35-mm) rocket. This 36-mm rocket is shorter and lighter than the LAW's 66-mm tactical rocket. It simulates the tactical rocket's smoke and flight trajectory but with less noise and backblast. The expended launcher is likely to wear out before the subcaliber inner tube does. When this happens, the inner tube can be transferred to another expended launcher. The M190 subcaliber launcher can be used in all training phases, from a fixed firing line to simulated tactical situations such as a squad live-fire exercise.

NOTE: The local Training Support Center (TSC) can install the conversion kits, but DOD Regulation 5100.76-M requires that all sensitive conventional arms, ammunitions, and explosives, including expended launchers, be carefully controlled.

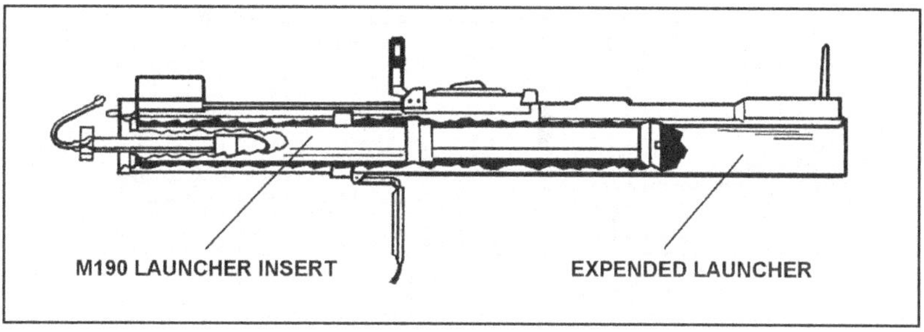

Figure B-1. LAW subcaliber trainer.

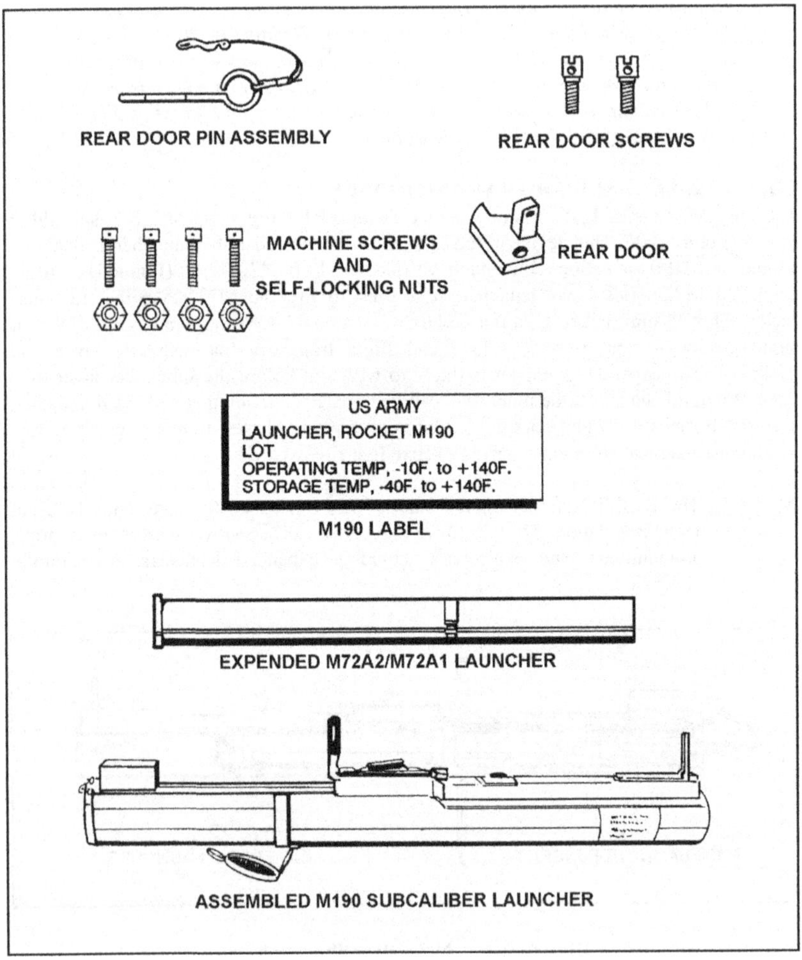

Figure B-2. M190 subcaliber conversion kit.

a. **Technical Data.**

(1) Technical data for the M73 35-mm subcaliber practice rocket are as follows:

Length .... 22.48 cm (8.86 inches)
Diameter .... 3.51 cm (1.38 inches)
Weight .... 154 grams (5.39 ounces)
Range .... 10 to 250 meters (9.14 to 228.5 yards)
Propellant Charge .... 3 tubular grains of M7
Spotting Head Charge .... 05 oz (1.5 grams) of composition mix M80

(2) Technical data for the assembled M190 practice rocket launcher are as follows:

Diameter of launcher .... 12.45 cm (4.9 inches)
Diameter of rocket tube .... 3.56 cm (1.4 inches)
Weight (with rocket) .... 2.26 kg (5.02 pounds)
Length of launcher
  Extended .... 89.9 cm (35.4 inches)
  Collapsed .... 64.26 cm (25.32 inches)
Sights .... Open, temperature compensating

b. **Trigger**. This is the same as for the LAW.

c. **Sights**. These are the same as for the LAW.

d. **Operation**. The LAW subcaliber training device operates the same as an M72-series LAW. After the M190 has been loaded, it may remain closed, or it can be extended and fired. Either way, the procedures are the same as those for the LAW.

CAUTION

When loading the law subcaliber trainer, ensure the rear cover is open, the sling assembly is off, and the launcher is collapsed.

e. **Loading**. To load the M73 rocket into the M190 launcher (Figure B-3)—

(1) Twist and remove the primer housing lockpin and open the primer housing door.

(2) Remove the fired primer block (if present) from the primer block cavity of the launcher.

(3) Inspect the primer block cavity for obstructions such as primer case stuck on the firing pin.

(4) Pivot and remove the safety clip from the rocket.

(5) Insert the M73 rocket into the launcher with the primer face up and toward the firing pin housing.

(6) Guide the primer block into the primer block cavity with the primer toward the front of the launcher. Close the primer housing door, insert the lockpin across the primer housing door, and twist to ensure a spring-tight fit. Ensure the primer housing door fits snugly.

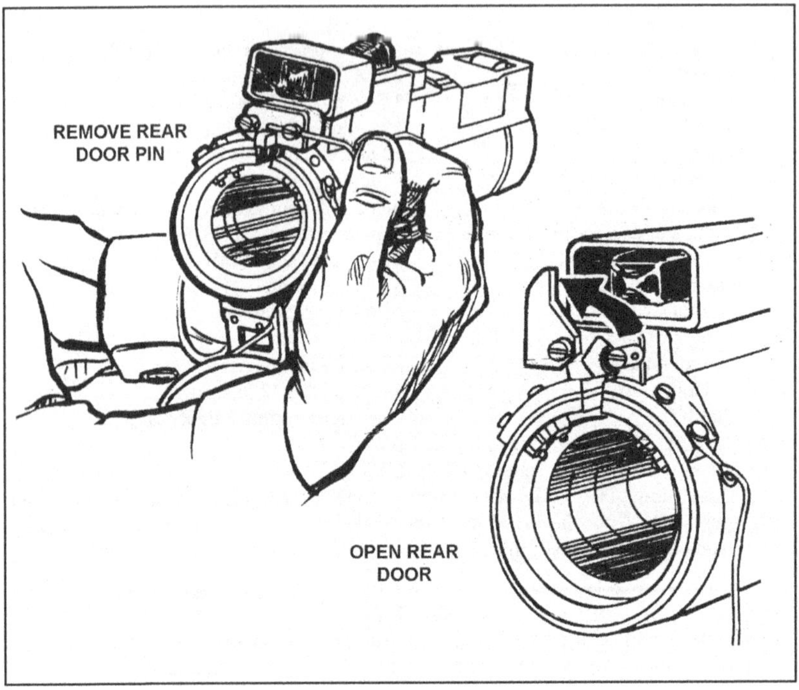

Figure B-3. Loading the LAW subcaliber launcher.

f. **Unloading**. To unload the M73 rocket from the M190 launcher—

(1) Return the arming handle to SAFE.

(2) Remove the weapon from your shoulder, keeping the weapon pointed downrange.

(3) Partly collapse the launcher, and remove the primer housing lockpin.

(4) Open the primary housing door, and remove the primer from its housing.

(5) Remove the rocket.

(6) Replace the safety clip on the rocket.

(7) Place the rocket in its original container.

g. **Function**. The M73 rocket launches the same as tactical rocket, but when the rocket head strikes the target, it sets off the spotting head that produces a flash, noise, and white smoke.

CAUTION

The M73 subcaliber rocket operates between –23 degrees and 60 degrees centigrade (-10 degrees and 140 degrees fahrenheit). Fire it only within this temperature range.

h. **Misfire Procedures**. These are the same as for the M-72 series LAW.

i. **Range Procedures and Safety**. These procedures are the same as for the LAW. The training device can be used against all solid stationary or moving targets. However, because the subcaliber rocket can penetrate 0.124 inches (0.315 cm) of steel plate or 8 inches (20.32 cm) of soft wood, the target should be constructed of 3/16-inch steel plate backed by 3/4-inch plywood.

j. **Combat Techniques**. These are the same as for the LAW.

B-2. AT4 FRONT SIGHT TEMPLATE AND TARGET SILHOUETTE SHEET

This graphic training aid (GTA 7-2-5), Sight Engagement Trainer, AT4, consists of two parts: Part 1, a front sight template, and Part 2, a target silhouette sheet. These are used together to train soldiers to obtain a correct sight picture. To obtain the correct sight picture for the target speed and range, the front sight template is placed on the target silhouette sheet as shown in Figure B-4. For each of the six targets, the reference number sheet shows the number that should appear in the circle on the front sight template. The reference number sheet is also used to determine the correct answer for any unit-developed test concerning Training Objective 12 task, Determine the Correct Sight Picture for the M136 (AT4). The local TSC can provide GTA 7-2-5.

Figure B-4. Using GTA-7-2-5 with target silhouette sheet and front sight template.

B-3. AT4 FIELD HANDLING TRAINER

The AT4 field handling trainer (FHT) is an inert device made from an expended AT4 tactical launcher. It simulates the weight, balance, characteristics, and operation of the launcher. The FHT is marked with either a gold or yellow 1-inch band between the front and rear sights, and with the word "DUMMY" in 1-inch letters on the side of the launch tube. The local TSC can convert expended launchers to training devices, but DOD Regulation 5100.76-M requires that these items be carefully controlled. The FHT is appropriate for several training situations that do not require the M287 tracer trainer, which prevents unnecessary damage to the tracer trainer. The FHT is used in marksmanship training, FTXs, and institutional training to help soldiers learn how to do the following:

- Inspect the AT4 for serviceability.
- Prepare an AT4 for firing.
- Demonstrate the correct AT4 firing position.
- Perform misfire procedures.
- Return an AT4 to carrying position.

B-4. AT4 SUBCALIBER TRACER TRAINER

The M287 subcaliber tracer trainer (Figure B-5) uses the 9-mm M939 training practice-tracer (TP-T) cartridge. When loaded, this trainer simulates the AT4 in weight, balance, and operation. The velocity and trajectory of its ammunition match that of the AT4's HEAT cartridge, but the M287 produces less noise, backblast, and overpressure. The M287 subcaliber tracer trainer is used in place of the AT4 in training.

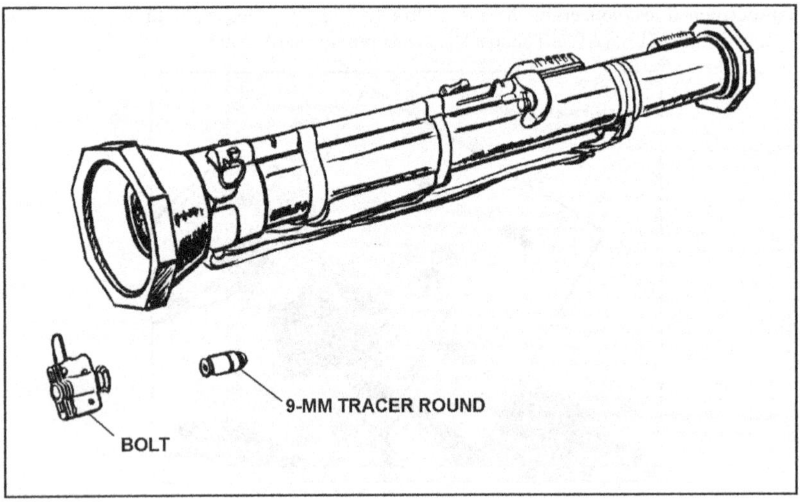

Figure B-5. M287 subcaliber tracer trainer.

a. **Description**. The M287 tracer trainer is a specially constructed AT4. Unlike the LAW subcaliber trainer, it is *not* made from an expended launcher, so it has its own NSN. It is designed to accept a special rifle barrel that fires a reduced-load 9-mm cartridge.

(1) The M287 may be fired at stationary or moving targets. Before it can be fired at a target tank occupied by personnel, the parts of the tank that could suffer damage must be shielded. The local TSC can provide specifications for modifying tanks to be used as targets for the M287 tracer trainer.

(2) Unlike the M136 AT4 itself and the FHT, the M287 has no band between the front and rear sights. The trainer has a 9-mm submachine gun barrel, a breach assembly, and a bolt (Figure B-6). The bolt is easily removed to load the 9-mm cartridges and to inspect the barrel for obstructions.

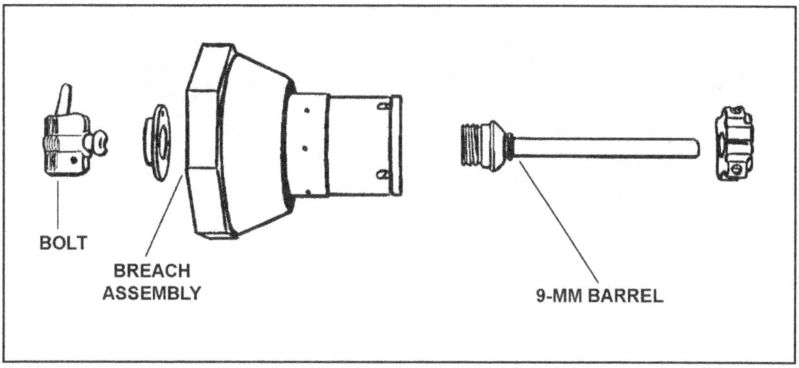

Figure B-6. M287 barrel assembly and components.

(3) If the M287 tracer trainer is damaged, the unit armorer can replace its complete firing mechanism, firing rod, bolt, sights, sling, and 9-mm barrel assembly (Figure B-7).

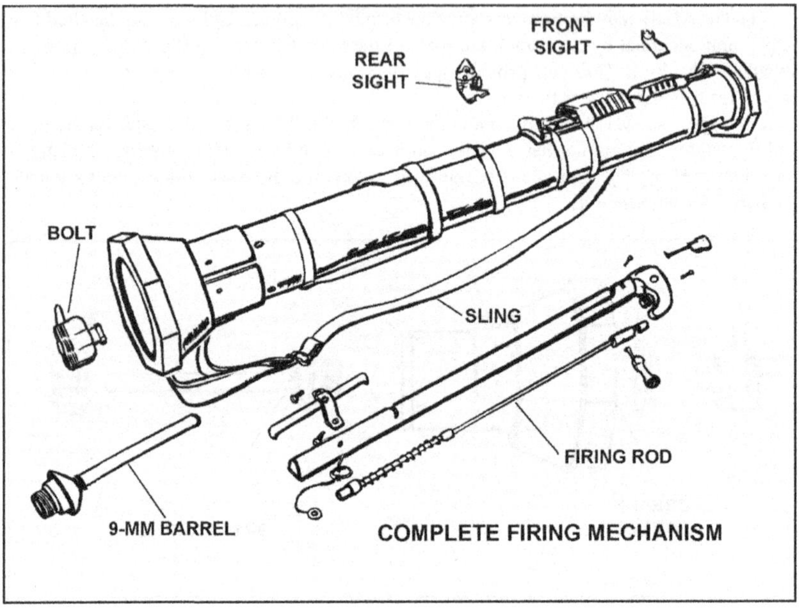

Figure B-7. Replaceable parts for the M287 tracer trainer.

b. **Technical Data**. The following technical data apply to the M287 subcaliber tracer trainer:

Length	1,020 mm (40 inches)
Weight	7 kg (15 pounds)
Action	Mechanical
Sights:	
Front	Three-post system
Rear	Adjustable range, with two peepholes: 2-mm peephole for daylight and 7-mm peephole for limited visibility
Operating temperature	10 degrees to 27 degrees Centigrade (0 degrees to 100 degrees Fahrenheit)
Muzzle velocity	300 mps (984 fps)
Caliber	9 mm

c. **Ammunition**. The M939 9-mm tracer cartridge has a lighter powder charge than a standard 9-mm bullet. The lighter charge enables the cartridge to closely duplicate the trajectory of the AT4 tactical round at ranges out to 700 meters. The M939 cartridge also has a tracer element to enable the firer to compare the impact of the cartridge with the sight picture. The firer can see the tracer out to 550 meters. Firing the M287 tracer trainer helps the firer learn the correct sight picture for moving targets. The cartridge's red tip and half-black base (Figure B-8) distinguishes it from standard 9-mm ammunition, which should never be fired from the M287 tracer trainer.

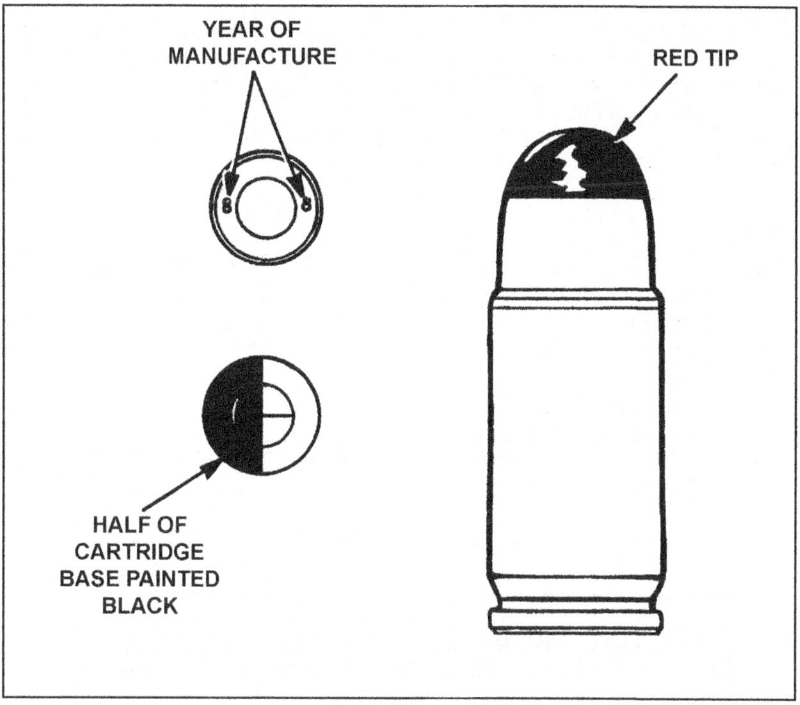

Figure B-8. M939, 9-mm tracer cartridge.

d. **Function Check**. Before the M287 is fired, a function check must be performed to ensure the trigger and safety mechanisms are operating properly. Function check procedures are shown in Table B-1. However, before performing a function check, ensure—

- The *cocking lever* is in the safe ("S") position.
- The *transport safety pin* is fully inserted, with the lanyard wrapped clockwise around the launcher.
- The *bolt* is removed from the breech.

STEP	OPERATOR ACTIONS	FUNCTION CHECK	CORRECTIVE MEASURES
1	Try to cock the tracer trainer	It should not cock	If it cocks, turn the trainer in for repair
2	Remove the transport safety pin and depress the forward safety	The forward safety should spring back when released	If it does not spring back, turn the tracer trainer in for repair
3	Cock the firing mechanism, pressing only the red trigger button	The firing rod should move only slightly and should be captured by the safety lever assembly. No more than half the rod should protrude through the rear of the firing assembly	If the tracer trainer fires, turn it in for repair
4	Recock the firing mechanism; fully depress and continue to hold down the forward safety; press the red trigger button	The tracer trainer should fire	If the tracer trainer *fails* to fire, turn it in for repair
5	Return the cocking lever to the SAFE position and reinstall the transport safety pin		

Table B-1. Function check, M287 tracer trainer.

WARNING
LOAD LIVE AMMUNITION ONLY ON THE FIRING LINE. NEVER FIRE THE M287 TRACER TRAINER AT HARD TARGETS LESS THAN 125 METERS FROM THE FIRING LINE. REMAIN CLEAR OF THE FRONT OF THE M287, WHICH MUST BE POINTED DOWNRANGE AT ALL TIMES.

e. **Loading**. The M287 tracer trainer is loaded as follows:

(1) The firer performs the function check and places the M287 tracer trainer on his shoulder.

(2) The trainer looks through the barrel from the rear to verify that it contains no obstructions. He inspects the cartridge primer to ensure it is not dented, inserts the bolt into the breech (Figure B-9), and presses in and turns the bolt clockwise to the "S" to

place it in the safe position (Figure B-10). Then he arms the tracer trainer by turning the bolt clockwise to "F" so that it can be fired (Figure B-11).

(3) With the launcher still on his shoulder, the firer cocks the launcher by placing the cocking lever in the cocked position.

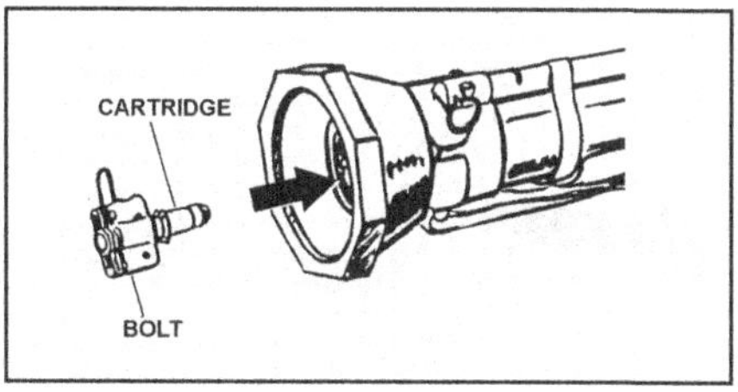

Figure B-9. Inserting bolt with cartridge into the breech of the M287 tracer trainer.

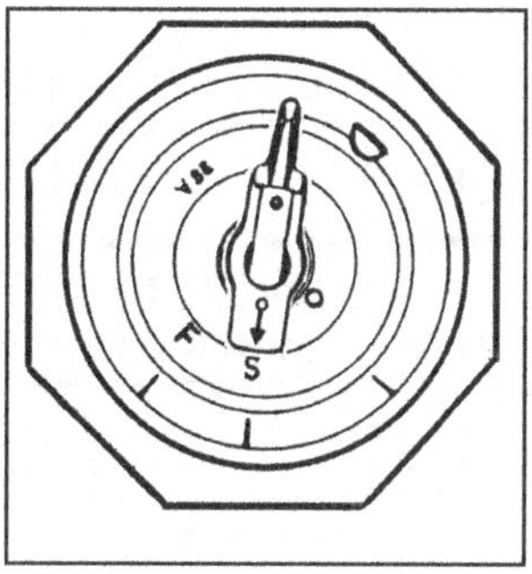

Figure B-10. Ensuring tracer trainer bolt is in the safe position.

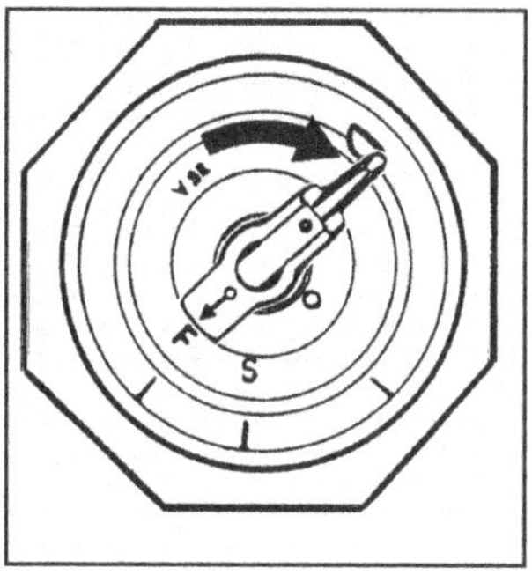

Figure B-11. Arming the tracer trainer.

f. **Unloading**. With the tracer trainer still on the firer's shoulder, the trainer or instructor removes the bolt by turning it counterclockwise past the "S," then pulling the bolt from the breech (Figure B-12). If more rounds are to be fired, he removes the expended cartridge and reloads the bolt.

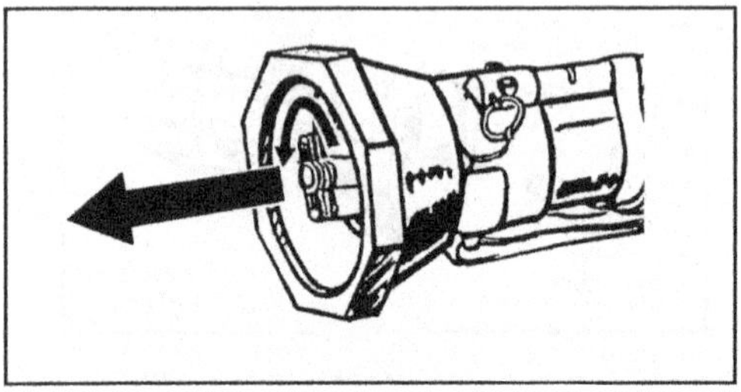

Figure B-12. Unloading the tracer trainer.

g. **Reloading**. With the tracer trainer still on his shoulder, the firer places the cocking lever in the safe position, keeping the tracer trainer on his shoulder and pointing toward the target. The trainer reloads the tracer trainer and prepares it for firing.

DANGER
NEVER TOUCH THE TRIGGER WHILE RELOADING.

h. **Misfire Procedures**. In the event of a misfire—

(1) Immediately shout "Misfire!"

(2) Release the forward safety, recock the launcher, and ensure the cocking lever is erect. Fully depress and hold down the forward safety, then aim and press the red trigger button.

(3) If the tracer trainer still fails to fire, repeat the two previous steps.

(4) If the launcher fails to fire again, release the forward safety, maintain your firing position, and return the cocking lever to the safe position.

(5) Notify the training supervisor.

(6) If a real misfire occurs that cannot be corrected by following the steps previously discussed, the trainer or assistant trainer on duty must ensure the cocking lever is in the safe position and that the forward safety is in the vertical position. He must then insert the transport safety pin and remove and inspect the cartridge. If the primer is dented, he replaces it, disposing of the old one IAW range safety SOP. If the primer is not dented, he notifies DS maintenance to inspect the bolt firing pin for damage. He then reloads the tracer trainer and continues the training exercise.

i. **Range Procedures and Safety**. These are the same as for the M136 AT4.

APPENDIX C

PERFORMANCE EVALUATIONS

The trainer administers performance evaluations to determine how much training the soldiers have learned. Those who fail are retrained and retested, and those who pass help retrain and evaluate those who did not. Instructions are given for right-handed firers: the M136 AT4 may only be fired right-handed, but the M72-series LAW may be fired left-handed, if procedures are reversed. Tasks are generic to both light antiarmor weapons. Conditions and standards may also be generic but, if not, are given by weapon.

TASK 1: Perform serviceability checks on a light antiarmor weapon.

CONDITIONS: Given an expended light antiarmor weapon under suitable conditions or in an appropriate location.

M72-Series LAW

STANDARDS: The firer performs the following actions:

TRAINING AND EVALUATION

1. Inspects the launcher to ensure it has no cracks, dents, or bulges.

2. Ensures the trigger arming handle is present and in the SAFE position.

3. Checks the rubber boots around the trigger bar and detent for cracks, tears, and deterioration.

4. (*M72A2 only*) Inspects the data plate on the right forward side of the launcher to ensure the words "with coupler" appear on the first line. If not, informs the trainer.

M136 AT4

STANDARDS: The firer performs the following actions:

TRAINING AND EVALUATION

1. Ensures the launcher has no cracks, dents, or bulges, or reports these defects to the instructor.

2. Ensures the transport safety pin is in place, or reports this defect to the instructor.

3. Ensures the cocking lever is in the SAFE (uncocked) position, or reports this defect to the instructor.

4. Ensures the muzzle cover is intact, or reports this defect to the instructor.

5. Ensures the color-coded band is correct, or reports this defect to the instructor.

6. Ensures the sights function properly, or reports this defect to the instructor.

7. Ensures the venturi has no foreign objects, or reports this defect to the instructor.

8. Ensures the forward safety functions properly, or reports this defect to the instructor.

9. Ensures the red trigger button is not broken or missing, or reports this defect to the instructor.

10. Ensures the shoulder stop is not broken or damaged, or reports this defect to the instructor.

11. Ensures the carrying sling is not frayed and is firmly attached to the launch tube, or reports this defect to the instructor.

	RESULTS	GO	NO-GO

Critique and grade the soldier.

TASK 2: Place a light antiarmor weapon into operation.

CONDITIONS: Given an expended light antiarmor weapon (FHT) on a range or in garrison.

M72-Series LAW

STANDARDS: The firer performs the following actions, in sequence:

TRAINING AND EVALUATION

1. Removes the rear cover pull pin.

2. Releases the rear cover and sling assembly.

3. Extends the launcher vigorously and ensures it is fully extended.

4. Places the launcher on his firing shoulder.

5. Checks the backblast area.

6. Pulls the arming handle out.

7. Sights.

M136 AT4

STANDARDS: The firer performs the following actions, in sequence:

TRAINING AND EVALUATION

1. Cradles the AT4 in his arms.

2. Removes the transport safety pin.

3. Unsnaps and unfolds the shoulder stop.

4. Places the AT4 on his shoulder.

5. Releases the sights.

6. Checks the backblast area.

7. Places cocking lever in the ARMED (cocked) position.

8. Fully depresses and holds down the forward safety.

RESULTS	**GO**	**NO-GO**
Critique and grade the soldier.		

TASK 3: Demonstrate the standing position used to fire a light antiarmor weapon.

CONDITIONS: Given a target and an expended light antiarmor weapon in any suitable location.

STANDARDS: The firer demonstrates the standing position as follows:

TRAINING AND EVALUATION

M72-Series LAW

1. Faces the target by executing a half-left face turn and spreading feet a comfortable distance apart.

2. Places the launcher on the shoulder with the left hand directly under the forward portion of the launcher.

3. Places the right elbow against the body for stability.

M136 AT4

1. Faces the target by executing a half-left face turn and spreads the feet a comfortable distance apart.

2. Places the launcher on the shoulder and grasps the sling near the launcher's muzzle with the left hand.

3. Places the right elbow against the body for stability.

	RESULTS	GO	NO-GO
Critique and grade the soldier.			

TASK 4: Demonstrate the kneeling position used to fire a light antiarmor weapon.

CONDITIONS: Given a target and an expended light antiarmor weapon in a suitable location, with the soldier in the standing position.

STANDARDS: The firer demonstrates the kneeling position as follows:

TRAINING AND EVALUATION

1. Kneels on the right knee with the right thigh at a 90-degree angle to the line of aim.

2. Sits back on the right heel and shifts weight forward.

3. Rests the upper left arm forward of the left knee.

4. Places the right arm against the body.

RESULTS **GO NO-GO**

Critique and grade the soldier.

TASK 5: Demonstrate the sitting position used to fire a light antiarmor weapon.

CONDITIONS: Given a light antiarmor weapon training device or expended launcher in a suitable location, with the soldier in the standing position.

STANDARDS: Each firer demonstrates the sitting position as follows:

TRAINING AND EVALUATION

1. Keeps the launcher pointed toward the target.

2. Executes a half-left face turn and sits with the legs crossed.

3. Places the launcher on the right shoulder.

4. Leans slightly forward from the hips.

5. Rests the elbows forward of the knees to avoid bone-to-bone contact.

RESULTS **GO NO-GO**

Critique and grade the soldier.

TASK 6: Demonstrate the prone position used to fire a light antiarmor weapon.

CONDITIONS: Given a target and an expended light antiarmor weapon (FHT) in a suitable location.

STANDARDS: The firer demonstrates the prone position as follows:

TRAINING AND EVALUATION

1. To keep clear of the backblast area, lies down at an angle of not less than 45 degrees from the line of fire.

2. Keeps back straight and right leg directly on line with the right hip and right shoulder.

3. Moves left leg as far as possible to ensure comfort.

4. Holds both elbows well below the launcher.

5. Checks the backblast area to ensure that no part of the body is in it.

6. Holds head as steady as possible, with the firing eye lined up with the sights.

RESULTS **GO** **NO-GO**

Critique and grade the soldier.

TASK 7: Identify enemy vehicles and weapons.

CONDITIONS: Given ten vehicle identification cards (GTA 17-2-13).

STANDARDS: The soldier correctly identifies eight out of ten vehicles or weapons by nomenclature.

RESULTS **GO** **NO-GO**

Critique and grade the soldier.

TASK 8: Estimate range.

CONDITIONS: Given stationary, partly exposed, or fully exposed targets at ranges between 50 and 250 meters, during daylight, in weather conditions where all targets are visible. Targets include personnel, equipment, silhouettes, and vehicles.

STANDARDS: The soldier states the distance to each target with no more than 20 percent error.

RESULTS **GO** **NO-GO**

Critique and grade the soldier.

TASK 9: Apply the correct lead for a target.

CONDITIONS: Given a sight template, a series of ten flank, frontal, and oblique silhouettes of armored targets, rates of movement, and ranges. (At least three different views, speeds, and ranges are given.)

STANDARDS: The soldier uses correct sighting procedures and correctly applies the lead eight out of ten times.

RESULTS **GO NO-GO**

Critique and grade the soldier.

TASK 10: Perform misfire procedures on a light antiarmor weapon.

CONDITIONS: Given a previously fired light antiarmor weapon and instructions that the weapon has misfired, on a suitable firing range, under simulated combat conditions, in any lighting conditions.

M72-series LAW

STANDARDS: The soldier applies misfire procedures as follows and tries to fire the M72-series LAW. If the weapon still fails to fire, he disposes of it IAW unit SOP:

TRAINING AND EVALUATION

1. Immediately squeezes the trigger bar again.

2. If the LAW still does not fire, tries to place the trigger arming handle on SAFE.

3. Removes the launcher from the shoulder and partly collapses it (10 to 15 centimeters or 4 to 6 inches), keeping the launcher pointed upward and downrange.

4. Extends the launcher.

5. If the LAW fails to fire again, squeezes the trigger, tries to return the weapon to SAFE, partly collapses the launcher, and sets the weapon aside.

M136 AT4

STANDARDS: The soldier performs combat misfire procedures on the AT4 as follows:

TRAINING AND EVALUATION

1. Releases the forward safety with the right hand.

2. Recocks the firing mechanism using the cocking lever.

3. Presses and holds down the forward safety, aims, and presses the trigger.

4. If launcher still fails to fire, repeats Steps 1 and 2.

5. If the launcher again fails to fire, returns the cocking lever to the SAFE (uncocked) position.

6. Removes the weapon from the shoulder, keeping it pointed toward the target.

7. Replaces the transport safety pin.

8. Informs the instructor that the firer breaks the sights off to indicate a misfired weapon when in combat.

9. Places the launcher on the ground, pointing toward the target.

RESULTS	GO	NO-GO

Critique and grade the soldier.

TASK 11: Return the light antiarmor weapon to its carrying configuration.

CONDITIONS: Given an expended light antiarmor weapon in a suitable location.

STANDARDS: The soldier performs the following actions, in sequence:

M72-series LAW

TRAINING AND EVALUATION

1. Pushes in the trigger arming handle.

2. Removes the launcher from the shoulder.

3. Depresses the barrel detent and collapses the launch tube, guiding the front and rear sights into position.

4. Replaces the sling assembly.

5. Replaces the rear cover pull pin.

M136 AT4

1. Releases the forward safety.

2. Returns the cocking lever to the SAFE (uncocked) position by pushing forward and upward on it, then letting it spring back into position.

3. Removes the launcher from the shoulder, ensuring the muzzle is pointed in the direction of fire.

4. Replaces the transport safety pin, fully seating it in the retainer hole.

5. When closing the rear sight, resets the range indicator to the 200-meter setting. (If this step is not accomplished, the rear sight breaks off when the sight cover is closed.)

6. Restores the sights by laying them down and closing their covers.

7. Snaps the shoulder stop into the closed position.

8. Moves to another location; slings the launcher over either shoulder and moves out.

RESULTS **GO NO-GO**

Critique and grade the soldier.

APPENDIX D

SUGGESTED DESIGN FOR COMBINED LIGHT ANTIARMOR RANGE

This appendix provides basic guidance for units to establish a range complex for light antiarmor weapons. The complex described may be used for multilevel training and firing of light antiarmor weapons; both subcaliber trainers and live HE ammunition may be used on the same range. Training areas should be near, but not adjacent to, the firing line. The firing line should be designed to allow personnel to observe firing from the side. The target array should include stationary targets composed of F-type targets or the hulls of hard targets (APCs or tanks), a moving target (a target on track or a manned moving target tank [MMTT]), and bunkers. This allows the firer to engage a variety of targets and to develop his skills in the various types of engagements.

D-1. FIRING LINE

The firing line should be organized to allow for the 100-meter AT4 backblast (Figure D-1). Also, at least 20 meters must be allowed between firing points. The backblast area must be fenced, roped, or marked in some way, so soldiers know not to enter it when firing is being conducted. A range that meets the requirements of the AT4 will also meet the requirements of the LAW.

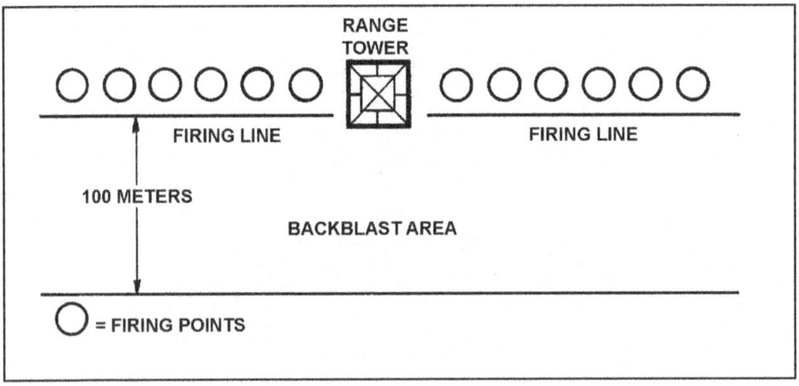

Figure D-1. Firing line.

D-2. TARGET ARRAY

The target array should include targets at ranges of 100 to 700 meters (Figure D-2). This enables the unit to conduct advanced instructional fire on targets located beyond the weapons' maximum effective ranges. Standard vehicle silhouettes or tank hulls should be used for stationary targets. Selected targets should be positioned behind mounds of earth to simulate vehicles in defilade or partly behind trees or other objects. Firers adjust their points of aim accordingly. Targets should also be positioned to allow firers to engage flank, frontal, and oblique targets. Bunker targets should be placed within the target array at ranges of 100 to 400 meters. Bunker targets should be no more than 1 meter high or 2 1/2 to 3 meters wide. Moving target silhouettes should travel along a track or road so the firers can engage fast moving targets from both flank and oblique angles. The moving targets should be placed at ranges of 150 to 600 meters and should enter and exit from the flanks outside the range fan. Firers may only use HE weapons to engage hard targets. This reduces the damage to other targets the unit must maintain such as F-type silhouettes and bunkers.

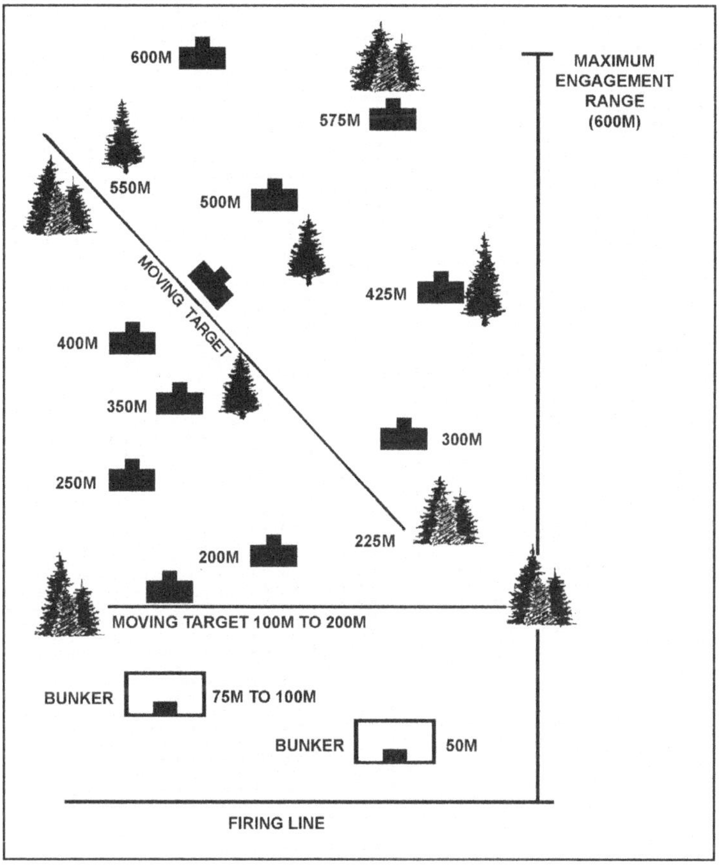

Figure D-2. Target array.

APPENDIX E

FIRING TABLES AND EXAMPLE SCORECARDS

This appendix provides firing tables for the subcaliber gunnery trainers discussed in this manual. It also includes an example of a completed scorecard for each light antiarmor weapon. All instructional and qualification firing tables are conducted on suitable ranges with ammunition designated in the firing tables. Some tables are fired in daylight, others in limited visibility, but all have some NBC firing (Table E-1).

M72-Series LAW	Instructional day fire, stationary targets
	Instructional day fire, moving targets
	Instructional night fire
M136 AT4	Instructional fire
	Instructional night fire
	Qualification fire
	Qualification night fire
	Advanced instructional fire

Table E-1. Firing tables for light antiarmor weapons.

E-1. M72-SERIES LAW

The M190 35-mm subcaliber device is used to fire the instructional firing tables for the M72-series LAW and DA Form 7323-R (M72-Series LAW Scorecard) is used to score them. Figure E-1 shows an example of the completed form. The instructional day and night firing tables for the LAW are provided in Tables E-2 and E-3.

NOTE: A blank copy of the form is located at the end of this publication for local reproduction on 8 1/2-inch by 11-inch paper.

M72-SERIES LAW SCORECARD

Instructions for using this form are provided in FM 23-25.
TRADOC is the proponent agency.

NAME THIELE, STEPHEN SSN 123-45-6789 DATE 19 APR 94

GRADE E-5 RATING 1st CLASS UNIT HHC 2/29 INF. REGT. WEATHER (circle one) CLEAR (RAIN) FOG SNOW

DATA REQUIRED BY THE PRIVACY ACT OF 1974

AUTHORITY: 10 USC 3012(g) / Executive Order 9397 **PRINCIPAL PURPOSE(S):** Record individual performance. **ROUTINE USES:** Evaluate individual proficiency and determine proficiency level. SSN is used for positive identification purposes only. **MANDATORY OR VOLUNTARY DISCLOSURE AND EFFECT ON INDIVIDUAL NOT PROVIDING INFORMATION:** Voluntary. Individuals not providing information cannot be rated/scored on a mass basis.

TABLE	TYPE OF FIRE	TYPE OF TARGET	ROUND	RANGE (M)	FIRING POSITION	HIT	MISS
1	Instructional DAY	Stationary	1	125 to 150	Standing	✓	
			2	150 to 175	Kneeling	✓	
			* 3	150 to 200	Sitting	✓	
			* 4	125 to 150	Prone	✓	
					Table Subtotals	4	0

FIRER'S SIGNATURE Stephen Thiele DATE 19 APR 94 SCORER'S SIGNATURE Benny C. Martin DATE 19 APR 94

TABLE	TYPE OF FIRE	TYPE OF TARGET	ROUND	RANGE (M)	FIRING POSITION	HIT	MISS
2	Instructional DAY	Moving	1	100 to 200	Standing	✓	
			2	100 to 200	Sitting	✓	
			* 3	100 to 200	Kneeling		✓
					Table Subtotals	2	1

FIRER'S SIGNATURE Stephen Thiele DATE 20 APR 94 SCORER'S SIGNATURE Benny C. Martin DATE 20 APR 94

TABLE	TYPE OF FIRE	TYPE OF TARGET	ROUND	RANGE (M)	FIRING POSITION	HIT	MISS
3	Instructional NIGHT	Stationary	1	100 to 150	Kneeling		✓
			2	100 to 150	Sitting	✓	
			* 3	100 to 150	Sitting	✓	
					Table Subtotals	2	1

FIRER'S SIGNATURE Stephen Thiele DATE 21 APR 94 SCORER'S SIGNATURE Benny C. Martin DATE 21 APR 94

* Firer must wear a protective mask when firing this round.

CONSOLIDATED SCORES

FIRER'S SIGNATURE Stephen Thiele DATE 21 APR 94 TOTAL MISSES 2

SCORER'S SIGNATURE Benny C. Martin DATE 21 APR 94 RANK SFC

RANGE OIC'S SIGNATURE Terry Anderson DATE 21 APR 94 RANK CPT

INSTRUCTOR NOTES	
ALL FIRING TABLES	• Do not announce the ranges to targets. • Fire all engagements in sequence.

DA Form 7323, JUL 94

Figure E-1. Example completed DA Form 7323-R.

TABLE 1 M72-Series LAW Instructional Day Fire, Stationary Targets	
TASK	Each soldier engages stationary armored targets with the M190 subcaliber device.
CONDITIONS	Given one M190 subcaliber device and four M73 rockets, on a suitable firing range with stationary targets at a range of 100 to 225 meters. Two targets are engaged in MOPP4.
STANDARD	The soldier fires four rockets at stationary targets and achieves at least three hits.
TABLE 2 **M72-Series LAW Instructional Day Fire,** **Moving Targets**	
TASK	Each soldier engages moving armored targets with the M190 subcaliber device.
CONDITIONS	Given one M190 subcaliber device, three M73 rockets, and targets moving at a rate of 8 to 24 kmph and at a range of 100 to 200 meters. Frontal and oblique-moving targets may be used when a manned, moving tank is available. One target is engaged in MOPP4.
STANDARD	The soldier fires three rockets at moving targets and achieves at least two hits.

Table E-2. Instructional day firing tables for the M72-series LAW.

TABLE 3 M72-Series LAW Instructional Night Firing Before you can use the AN/PVS-4, you must install an M72A1 sight reticle on it and Boresight it to an expended M72-series LAW launcher.	
TASK	Each soldier engages stationary targets at night with the M190 subcaliber device.
CONDITIONS	Given one M190 subcaliber device and three M73 rockets, on a suitable firing range with stationary targets at a range of 100 to 150 meters, at night, and either an AN/PVS-4 or illumination provided by indirect fire.
STANDARD	The soldier fires three rockets at stationary targets and achieves at least one hit. He engages one of these targets in MOPP4.

Table E-3. Instructional night firing table for the M72-series LAW.

E-2. M136 AT4

The M287 9-mm tracer bullet trainer is used to fire the instructional, qualification, and advanced instructional firing tables for the M136 AT4 and DA Form 7324-R (M136 AT4 Scorecard) is used to score them. A copy of this form is located at the end of this publication for local reproduction on 8 1/2-inch by 11-inch paper. Figure E-2 shows an example of the completed form; Tables E-4 through E-6, on pages E-7, E-8, and E-9, provide the instructional, qualification, and advanced instructional firing tables for the AT4. The latter must be fired on a separate, more specialized range.

E-3. NIGHT COURSE OF FIRE TABLES AND QUALIFICATION FOR M136 AT4

Each soldier is given a PMI before the instructional and qualification tables are fired. Once the soldier completes Table I and III (day tables), he is given additional instruction on the use of the aided vision devices for the night operations.

a. **Qualification Tables for the M287 Tracer Trainer.**

- Table I (Instructional Day). This scenario consists of 12 targets from 100 to 300 meters. The standard to proceed to the next table is 6 out of 12 targets successfully engaged.
- Table II (Instructional Night). This scenario consists of 8 targets from 100 to 300 meters. The standard to proceed to the next table is 6 out of 12 targets successfully engaged at night.
- Table III (Qualification Day). This scenario consists of 8 targets from 100 to 300 meters. The standard to proceed to the next table is 6 out of 8 targets successfully engaged.
- Table IV (Qualification Night). This scenario consists of 8 targets from 100 to 300 meters (same as Table III, but conducted at night). The standard to proceed to the advanced firing tables is 6 out of 8 targets successfully engaged.

NOTES: 1. In Table IV, eight engagements are proposed as opposed to the current six engagements.
2. All references to the M136 launcher and cartridge refer to the M287 9-mm tracer bullet training device.

b. **Ammunition Requirements for Night Training Strategy**. The current STRAC ammunition allocation and course of fire is compared with the proposed night training strategy in Table E-7 on page E-9. The current STRAC table stipulates that 120 rounds of 9-mm tracer ammunition is authorized to each squad for additional squad members to fire Tables I and IV.

c. **Day Standards**. Commanders are able to achieve the current day standard at night with the M136 AT4 through the use of training strategy within this field manual. A gunner who is qualified on the day standard and has executed preliminary marksmanship training with the appropriate night vision device can achieve the current day standard at night. During testing, gunners had no problem engaging targets out to 300 meters. Table E-8, on page E-9, shows the day standard and is also the proposed night standard.

M136 AT4 SCORECARD

Instructions for using this form are provided in FM 23-25.
TRADOC is the proponent agency.

NAME DINKENS, DAVID SSN 135-24-5678 DATE 19 APR 94

GRADE E-3 RATING ______ UNIT HHC 2/29 INF REGT. WEATHER (circle one) CLEAR (RAIN) FOG SNOW

DATA REQUIRED BY THE PRIVACY ACT OF 1974

AUTHORITY: 10 USC 3012(g) / Executive Order 9397. **PRINCIPAL PURPOSE(S):** Record individual performance. **ROUTINE USES:** Evaluate individual proficiency and determine proficiency level. SSN is used for positive identification purposes only. **MANDATORY OR VOLUNTARY DISCLOSURE AND EFFECT ON INDIVIDUAL NOT PROVIDING INFORMATION:** Voluntary. Individuals not providing information cannot be rated/scored on a mass basis.

TABLE	TYPE OF FIRE	TYPE OF TARGET	ROUND	RANGE (M)	FIRING POSITION	HIT	MISS
		Stationary	1	150 to 200	Kneeling	✓	
		Stationary	* 2	150 to 200	Kneeling	✓	
		Stationary	3	250 to 300	Sitting	✓	
		Stationary	4	250 to 300	Sitting	✓	
1	Instructional Day	Stationary	5	125 to 150	Prone	✓	
		Stationary	6	125 to 150	Prone	✓	
		MMTT/MAS	7	150 to 200	Kneeling	✓	
		MMTT/MAS	* 8	150 to 200	Kneeling	✓	
		MMTT/MAS	9	200 to 250	Standing		✓
		MMTT/MAS	*10	200 to 250	Fighting position, unsupported		✓
		MMTT/MAS	11	250 to 300	Kneeling	✓	
		MMTT/MAS	12	250 to 300	Kneeling		✓
					Table Subtotals	9	3

FIRER'S SIGNATURE David Dinkens DATE 19 APR 94 SCORER'S SIGNATURE Dan Sell DATE 19 APR 94

TABLE	TYPE OF FIRE	TYPE OF TARGET	ROUND	RANGE (M)	FIRING POSITION	HIT	MISS
		Stationary	1	125 to 150	Prone		✓
		Stationary	2	125 to 150	Prone	✓	
		Stationary	* 3	150 to 200	Sitting	✓	
		Stationary	4	150 to 200	Sitting	✓	
		Stationary	5	250 to 300	Kneeling		✓
2	Instructional Night	Stationary	6	250 to 300	Kneeling	✓	
		MMTT/MAS	7	125 to 150	Standing	✓	
		MMTT/MAS	8	125 to 150	Standing	✓	
		MMTT/MAS	* 9	200 to 250	Kneeling	✓	
		MMTT/MAS	10	200 to 250	Kneeling	✓	
		MMTT/MAS	11	150 to 200	Standing		✓
		MMTT/MAS	12	150 to 200	Fighting position, unsupported	✓	
					Table Subtotals	9	3

FIRER'S SIGNATURE David Dinkens DATE 19 APR 94 SCORER'S SIGNATURE Dan Sell DATE 19 Apr 94

* Firer must wear a protective mask when firing this round.

INSTRUCTOR NOTES	
ALL FIRING TABLES	• Do not announce the ranges to targets. • Fire all engagements in sequence. • Fire Table 1 before Table 3; fire Table 2 before Table 4.
FIRING TABLES 2 THROUGH 5	• Ensure no one fires at front or rear of a manned moving tank target. • Total hits and misses, and enter results in the consolidated scores block on the back of this scorecard.

DA Form 7324-R, JUL 94 (FRONT)

Figure E-2. Example completed DA Form 7324-R (front).

M136 AT4 SCORECARD

TABLE	TYPE OF FIRE	TYPE OF TARGET	ROUND	RANGE (M)	FIRING POSITION	HIT	MISS
3	Qualification Day	Stationary	1	125 to 150	Standing	✓	—
		Stationary	2	150 to 200	Kneeling	✓	—
		Stationary	* 3	200 to 250	Sitting	✓	—
		Stationary	4	250 to 300	Sitting	✓	—
		MMTT/MAS	5	125 to 150	Kneeling	✓	—
		MMTT/MAS	6	150 to 200	Fighting position, unsupported	—	✓
		MMTT/MAS	* 7	200 to 250	Sitting	✓	—
		MMTT/MAS	8	250 to 300	Kneeling	✓	—
					Table Subtotals	7	1

FIRER'S SIGNATURE David Dinkens DATE 21 Apr 94 SCORER'S SIGNATURE Dan Sell DATE 21 Apr 94

TABLE	TYPE OF FIRE	TYPE OF TARGET	ROUND	RANGE (M)	FIRING POSITION	HIT	MISS
4	Qualification Night	Stationary	* 1	125 to 150	Standing or Prone	✓	—
		Stationary	2	200 to 250	Kneeling	✓	—
		Stationary	3	250 to 300	Sitting	✓	—
		MMTT/MAS	4	150 to 200	Sitting	✓	—
		MMTT/MAS	* 5	125 to 150	Kneeling	—	✓
		MMTT/MAS	6	200 to 250	Fighting position, unsupported	✓	—
					Table Subtotals	5	1

FIRER'S SIGNATURE David Dinkens DATE 21 Apr 94 SCORER'S SIGNATURE Dan Sell DATE 21 Apr 94

TABLE	TYPE OF FIRE	TYPE OF TARGET	ROUND	RANGE (M)	FIRING POSITION	HIT	MISS
5	Advanced Instructional	Sequence	1		Sitting	✓	—
		Sequence	2		Sitting	✓	—
		Pair	3		Kneeling	✓	—
		Pair	4		Kneeling	✓	—
		Volley	5		Kneeling	✓	—
		Volley	* 6		Sitting	✓	—
					Table Subtotals	6	0

FIRER'S SIGNATURE David Dinkens DATE 22 Apr 94 SCORER'S SIGNATURE Dan Sell DATE 22 Apr 94

* Firer must wear a protective mask when firing this round.

QUALIFICATION CRITERIA — BASED ON TOTAL HITS FOR TABLES 3 AND 4

FIRST CLASS .... 13 HITS
SECOND CLASS .... 11 HITS (circled)
THIRD CLASS .... 9 HITS
UNQUALIFIED .... FEWER THAN 9 HITS

CONSOLIDATED SCORES

INSTRUCTIONAL TOTALS (TABLES 1 AND 2): HITS 18 MISSES 6 CLASS RATING, TABLES 1 AND 2: FIRST

QUALIFICATION TOTALS (TABLES 3 AND 4): HITS 12 MISSES 2 OVERALL TOTALS HITS 30 MISSES 8

FIRER'S SIGNATURE David Dinkens DATE 22 Apr 94 SCORER'S SIGNATURE Dan Sell DATE 22 Apr 94

RANGE OIC'S SIGNATURE Benny Utley RANK SFC DATE 22 Apr 94

DA Form 7324-R, JUL 94 (BACK)

Figure E-2. Example completed DA Form 7324-R (back).

TABLE 1 M136 AT4 Instructional Fire	
TASK	Each soldier engages an armored target with the M287 tracer bullet trainer.
CONDITIONS	Given one M287 9-mm subcaliber device and twelve rounds of 9-mm tracer ammunition, a suitable firing range, stationary targets at a range of 100 to 300 meters, and other targets moving 8 to 24 kmph at a range of 150 to 200 meters. When a manned, moving tank is available, frontal and oblique-moving targets may be used.
STANDARD	The soldier fires twelve tracer bullets at stationary and moving targets. He engages three of these targets in MOPP4.
TABLE 3 M136 AT4 Instructional Fire	
TASK	Each soldier engages an armored target with the M287 tracer bullet trainer at night.
CONDITIONS	At night, given one M287 9-mm subcaliber device and twelve rounds of 9-mm tracer ammunition, artificial illumination, a suitable firing range, stationary targets at a range of 100 to 300 meters, and other targets moving from 8 to 24 kmph (5 to 15 mph) at a range of 150 to 200 meters. When a manned, moving tank is available, frontal and oblique-moving targets may be used.
STANDARD	The soldier fires twelve tracer bullets at stationary and moving targets. He engages two of these targets in MOPP4.

Table E-4. Instructional firing tables for the M136 AT4.

TABLE 2 M136 AT4 Qualification Fire	
TASK	Each soldier engages an armored target with the M287 tracer bullet trainer.
CONDITIONS	Given one M287 9-mm subcaliber device and eight rounds of 9-mm tracer ammunition, a suitable firing range, stationary targets at a range of 100 to 300 meters, and other targets moving at a rate of 8 to 24 kmph (5 to 15 mph) at a range of 150 to 200 meters. When a manned, moving tank is available, frontal and oblique-moving targets may be used.
STANDARD	The soldier fires eight tracer bullets at stationary and moving targets and achieves at least six hits. He engages two of these targets in MOPP4.
TABLE 4 M136 AT4 Qualification Fire	
TASK	Each soldier engages an armored target with the M287 tracer bullet trainer.
CONDITIONS	At night, given one M287 9-mm subcaliber device and six rounds of 9-mm tracer ammunition, artificial illumination, a suitable firing range, stationary targets at a range of 100 to 300 meters, and other targets moving 8 to 24 kmph (5 to 15 mph) at a range of 150 to 200 meters. When a manned, moving tank is available, frontal and oblique-moving targets may be used.
STANDARD	At night, given one M287 9-mm subcaliber device and six rounds of 9-mm tracer ammunition, artificial illumination, a suitable firing range, stationary targets at a range of 100 to 300 meters, and other targets moving 8 to 24 kmph (5 to 15 mph) at a range of 150 to 200 meters. When a manned, moving tank is available, frontal and oblique-moving targets may be used.

Table E-5. Qualification firing tables for the M136 AT4.

TABLE 5 M136 AT4 Advanced Instructional Fire Fire this table on an appropriate separate, specialized range.	
TASK	Each soldier engages an armored target with the M287 tracer bullet trainer.
CONDITIONS	Given one M287 9-mm subcaliber device and six rounds of 9-mm tracer ammunition, a suitable firing range, stationary targets at a range of 100 to 300 meters, and other targets moving 8 to 24 kmph (5 to 15 mph) at a range of 150 to 200 meters. When a manned, moving tank is available, frontal and oblique-moving targets may be used.
STANDARD	The soldier must fire six tracer bullets at stationary and moving targets and achieve at least three hits. He must engage two of the targets with sequence firing, two with pair firing, and two with volley firing, and he must observe MOPP4 while engaging one of the six.

Table E-6. Advanced instructional firing table for the M136 AT4.

EVENT	CURRENT STRAC	RECOMMENDED CHANGES	TOTAL
TABLE I	12	0	12
TABLE II	12	0	12
TABLE III (QUALIFICATION)	8	0	8
TABLE IV (NIGHT QUALIFICATION)	6	2	8
TABLE V (ADVANCE)	6	0	0
TOTAL	44	2	40

Table E-7. Ammunition requirements for night training strategy.

M136 AT4 (M287 TRAINER) TABLE III/IV			
ENGAGEMENT	RANGE	POSITION	HIT/MISS
1	125 to 150	Standing	
2	150 to 200	Kneeling	
3	200 to 250	Sitting	
4	250 to 300	Sitting	
5	125 to 150	Kneeling	
6	150 to 200	Fighting Position	
7	200 to 250	Sitting	
8	250 to 300	Kneeling	

Table E-8. M136 AT4 (M287 trainer) Table III/IV.

APPENDIX F

INFRARED LASER AIMING DEVICES

The newest infrared laser pointers increase the night firing accuracy of infantry weapons. The new family of weapon mounts (some fielded with the thermal weapon sight) and helmet brackets are more stable and reliable for continued accurate firing under field conditions. Their effectiveness is limited by the capability of the image-intensifying (I^2) sight with which they are used. Normally this is a night vision goggle or monocular, with or without the 3X magnifier. Again, they primarily enhance the gunner's ability to align his sights on the target, but usually have dramatic affect on his firing positions and scanning technique. Unlike I^2 sights mounted on the weapon, the gunner can somewhat reduce the adverse effects of muzzle flash by rising up higher off the sight with his helmet-mounted goggle or monocular. With the AN/PEQ-2A, the range of these devices is greatly improved, allowing great effectiveness for crew-served weapons out to 2,200 meters. However, trainers and leaders must adopt new safety procedures to ensure these lasers, which are not safe for the eyes, are treated like loaded weapons any time the batteries are issued. Finally, the bore light is also effective to boresight these pointers. Generally weapons do not require zeroing with live ammunition to confirm alignment after soldiers and leaders are qualified in the boresighting techniques. The AT4 cannot be conventionally boresighed, therefore a sight alignment method is used to zero the aided vision device.

F-1. AN/PAQ-4C, AIMING LIGHT

The AN/PAQ-4C aiming light (Figure F-1) projects an infrared laser beam, which cannot be seen with the eye but can be seen with night vision devices. This aiming light works with the AN/PVS-7B/C/D-series goggles and AN/PVS-14. The AN/PAQ-4C mounts on various weapons with mounting brackets and adapters.

a. **Technical Data**. The following technical data apply to the AN/PAQ-4C:

Optics:	100% parallax free, anti-reflective coated lens system
Length (sight):	14 centimeters (5.5 inches)
Weight:	164 grams (5.78 ounces)
Height:	3 centimeters (1.2 inches)
Width:	6.5 centimeters (2.5 inches)
Ranged:	Beyond 600 meters. (Actual range depends on light level and night vision device used for observation.)
Battery life:	100-hour operating (ON) time for AA batteries in temperatures above 0 degrees Centigrade (32 degrees Fahrenheit); 36 hours for temperatures below 0 degrees Centigrade (32 degrees Fahrenheit)

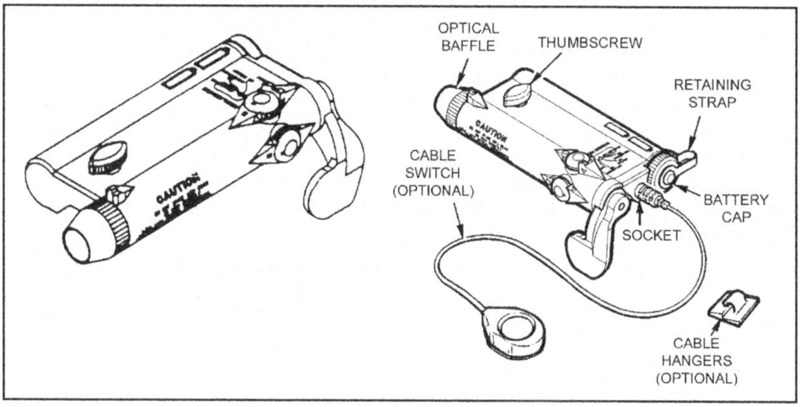

Figure F-1. AN/PAQ-4C, aiming light.

b. **Operation.** Activate the AN/PAQ-4C aiming light by rotating the ON/OFF switch lever or the button on the optional cable switch. Either switch connects power from two AA batteries to an internal electronic circuit, which produces the infrared laser. Internal lenses focus the infrared light into a narrow beam. Control the direction of the beam by rotating the mechanical adjusters with click detents. These adjusters are used to zero the aiming light to the weapon. Once zeroed to the weapon, the aiming light projects the beam along the line of fire of the weapon. The optical baffle prevents off-axis viewing of the aiming light beam by the enemy.

(1) ***M136 (AT4) Launcher and Cartridge Mounting Procedures***. The aiming light is attached to the M136 (AT4) launcher and cartridge by first attaching the mounting bracket to the weapon and then attaching the aiming light to the mounting bracket (Figures F-2 and F3).

STEP 1: Install the mounting bracket.

> **WARNING**
>
> Only M136 (AT4) qualified personnel should install and use the mounting bracket assembly. Conduct all M136 (AT4) operator preventive maintenance checks and services before installation of mounting bracket assembly.

NOTE: All references to the M136 (AT4) launcher and cartridge are also applicable to the M287 9-mm tracer bullet training device.

(a) Cradle the M136 (AT4) in left arm.

(b) Position the support bracket with the mounting rail on the left side and the marking FRONT over the rear sight.

(c) With the pivot bracket spread open, place the support bracket against the base of the rear sight housing and the bottom on the shoulder strap boss.

(d) Swing the pivot bracket around the M136 (AT4) and secure it by rotating the locking latch clockwise to engage the latch shaft.

(e) The lever screw assembly must be located in the rear threaded screw hole when mounting the aiming light. Lever screw assembly may require relocation from the front to the rear threaded screw hole.

(f) Place the bracket adapter (Figure F-2) in groove of the mounting rail so that the threaded screw hole in the base of the adapter is aligned with the lever screw assembly and tighten the lever screw.

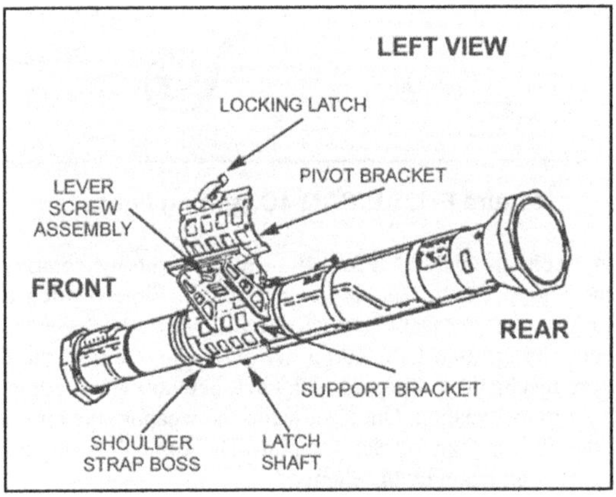

Figure F-2. M136 (AT4) installation.

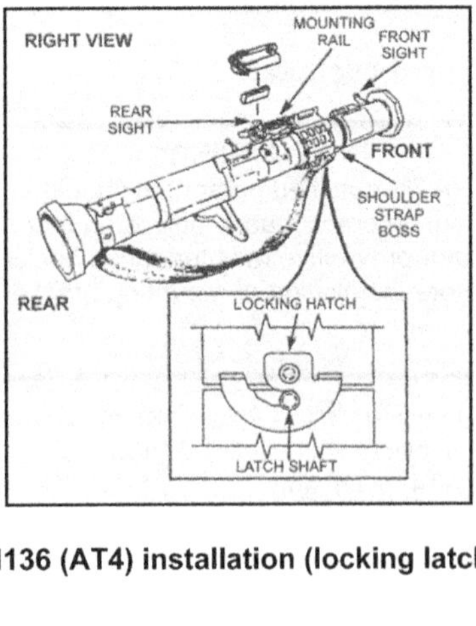

Figure F-3. M136 (AT4) installation (locking latch and shaft).

STEP 2: Install the bracket adapter (Figure F-4). The bracket adapter provides compatibility between the aiming light and the mounting brackets used on the M2, M60, and M136 AT4 weapons.

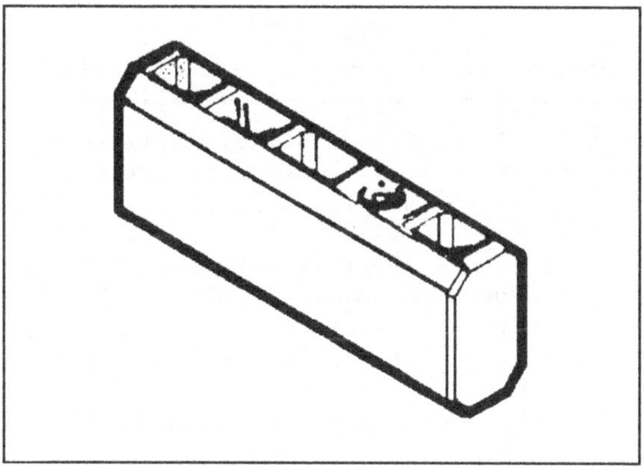

Figure F-4. Bracket adapter.

STEP 3: Attach the AN/PAQ-4 onto the bracket adapter (Figure F-5).

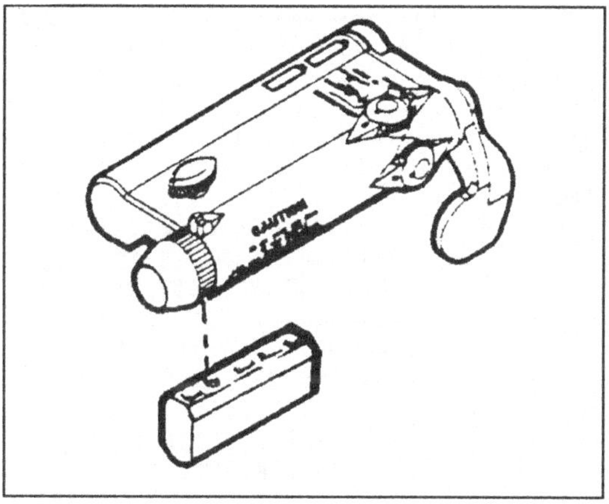

Figure F-5. Attaching bracket adapter to aiming light.

STEP 4: Attach the AN/PAQ-4 to the AN/PVS-4 mounting bracket.

- Position the AN/PAQ-4 on top of the bracket so that the spacer of the AN/PAQ-4 is aligned with the mounting knob of the bracket.
- Turn the mounting knob clockwise until the AN/PAQ-4 is tight.

WARNING
Do not store the AN/PAQ-4 with batteries installed.

(2) ***M136 (AT4) Launcher and Cartridge Dismounting Procedures***.

STEP 1: Detach the AN/PAQ-4 from the AN/PVS-4 mounting bracket.

- Turn the mounting knob counterclockwise until the AN/PAQ-4 is loose.
- Remove the AN/PAQ-4 from the bracket adapter.

STEP 2: Remove the adapter from the AN/PAQ-4

STEP 3: Remove the mounting bracket.

- Rotate the locking latch counter clockwise to disengage the latch shaft.
- Swing pivot bracket away from around the AT4.

F-2. AN/PEQ-2A TARGET POINTER/ILLUMINATOR/AIMING LIGHT

The AN/PEQ-2A target pointer/illuminator/aiming light (TPIAL) (Figure F-6) is a Class IIIb laser that emits a collimated beam of infrared light for precise aiming of the weapon as well as a separate infrared illuminating beam with adjustable focus. A safety block is provided for training purposes (blue side), which limits the operator from selecting the high power modes (black side). The TPIAL projects an infrared laser beam that cannot be seen with the eye but can be seen with night vision devices. It is also capable of projecting a much wider infrared illuminating beam from an integral illuminator. The TPIAL works with night vision goggles and mounts on various weapons with mounting brackets and adapters. The AN/PEQ-2A can also be used in the hand-held mode to illuminate and designate targets by leaders.

a. **Technical Data**.

Weight:	7.5 ounce (with 2 AA batteries)
Length:	16.26 centimeters (6.4 inches)
Width:	7.12 centimeters (2.8 inches)
Height:	3.05 centimeters (1.2 inches)
Range:	600 meters in low power (eye safe)
	2,000 meters in high power (non-eye safe)
Output Power:	
Aiming laser	25 m W (+- 10%)
Illuminator	30 m W (+50,-20%)
Beam Divergence:	
Aiming laser	.5 m rad
Illuminator	1.0-115 m rad
Batteries:	2 x 1.5V AA batteries

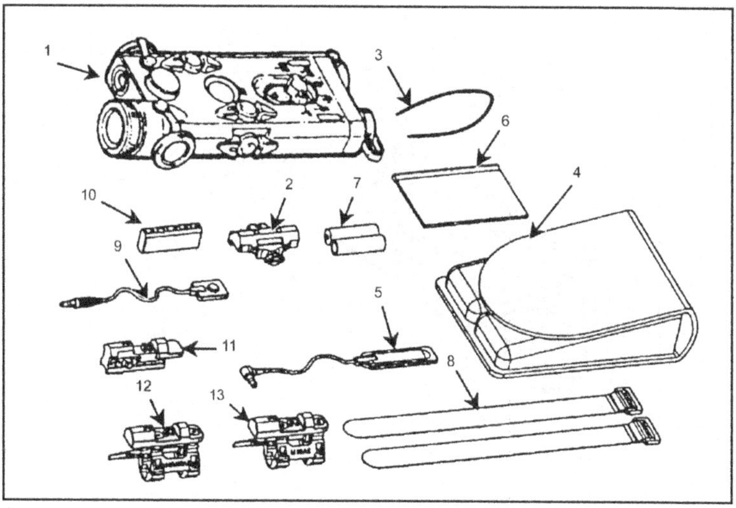

(1) TPIAL ASSEMBLY WITH SAFETY BLOCK
(2) RAIL
(3) NECK CORD
(4) TEXTILE BAG
(5) CABLE SWITCH, 12-INCH, MEMBRANE
(6) OPERATOR'S MANUAL
(7) BATTERIES 1.5V AA
(8) STRAP, RETENTION
(9) CABLE SWITCH, 20-INCH, BUTTON
(10) BRACKET ADAPTER
(11) TRAINING EXTENDER (ARMY ONLY)
(12) M4/M16A2 BRACKET ASSEMBLY

Figure F-6. AN/PEQ-2A with accessories.

b. **Operation.** This paragraph describes the battery installation, the safety block installation, the mode switch, the button switch, the cable switch, the focus switch, the lens cap, and the boresight adjusters.

(1) ***Battery Installation*** (Figure F-7).

(a) Unscrew the battery caps and install two AA batteries.

(b) Orient the batteries as indicated by the markings on the AN/PEQ-2A body.

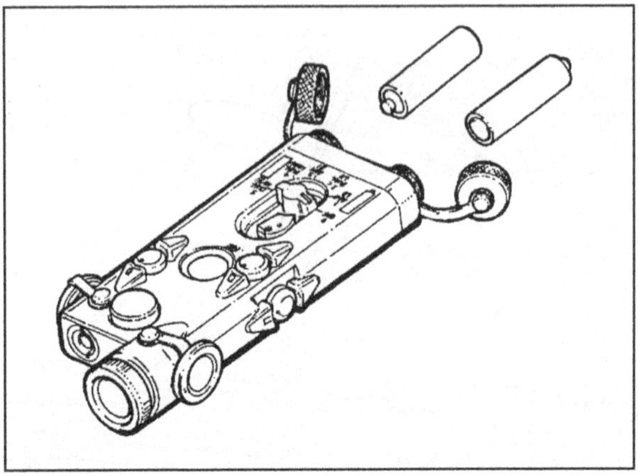

Figure F-7. AN/PEQ-2A battery installation.

(2) ***Safety Block Installation*** (Figure F-8). The safety block installed in the training mode (blue side up) prevents the operator from accessing the noneye safe modes (AIM HI, DUAL LO/HI, DUAL HI/HI). A .050 hex head Allen wrench is needed to unscrew the block from the body and reinstall it in the tactical mode (black side up).

Figure F-8. Safety block installation.

(3) ***Mode Selector***. The mode selector is used to set the mode in which the AN/PEQ-2A operates when the cable switch button or push button are depressed. The mode selector has six positions (Table F-1).

KNOB POSITION	OPERATION
0 OFF	The AN/PEQ-2A will not operate.
1 AIM LO	The aiming beam operates at low power.
2 DUAL LO	The aiming beam operates at low power and the illuminating beam operates at low power.
3 AIM HI	The aiming beam operates at high power.
4 DUAL LO/HI	The aiming beam operates at low power and the illuminating beam operates at full power.
5 DUAL HI/HI	The aiming beam operates at high power and the illuminating beam operates at full power.

Table F-1. Mode selector positions.

(4) ***Button Switch*** (Figure F-9). The button switch is used when the AN/PEQ-2A is hand held. Pressing the button switch operates the AN/PEQ-2A in the operational mode set by the mode selector. When the button is released, the AN/PEQ-2A turns off.

(a) A green light emitting diode (LED) is incorporated into the body of the AN/PEQ-2A to indicate that the AN/PEQ-2A is ON. Whenever the AN/PEQ-2A is activated, the green LED will light and stay lit until the unit is turned OFF.

(b) If continuous operation of the AN/PEQ-2A is desired, pressing the button switch twice in rapid succession will latch the AN/PEQ-2A ON. The AN/PEQ-2A will remain on until the push button is pressed a third time.

Figure F-9. Operation of the button switch.

(5) ***Cable Switch*** (Figure F-10). The cable switch is used when the AN/PEQ-2A is mounted on a weapon. The cable switch plugs into the back of the AN/PEQ-2A

assembly. Pressing the button or pad at the end of the cable switch causes the AN/PEQ-2A to turn on in the operational mode selected by the mode select switch. When the button is released, the AN/PEQ-2A turns off.

(a) If continuous operation of the AN/PEQ-2A is desired, pressing the cable switch twice in rapid succession will latch the AN/PEQ-2A ON. The AN/PEQ-2A will remain on until the push button is pressed a third time.

(b) When the cable switch plug is installed in the AN/PEQ-2A, it automatically locks into place. To remove the switch, pull back on the plug sleeve and pull the plug out. DO NOT TRY TO REMOVE THE PLUG BY PULLING ON THE CABLE.

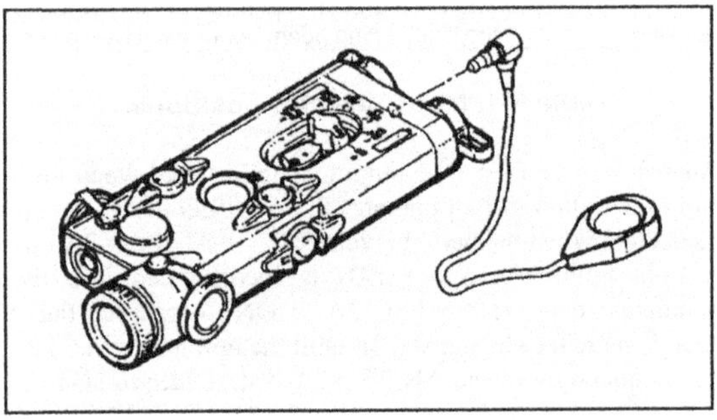

Figure F-10. Installation of the cable switch.

(6) ***Focus Knob*** (Figure F-11). The focus knob is used to vary the spread of the illumination beam based on the range and size of the area to be illuminated.

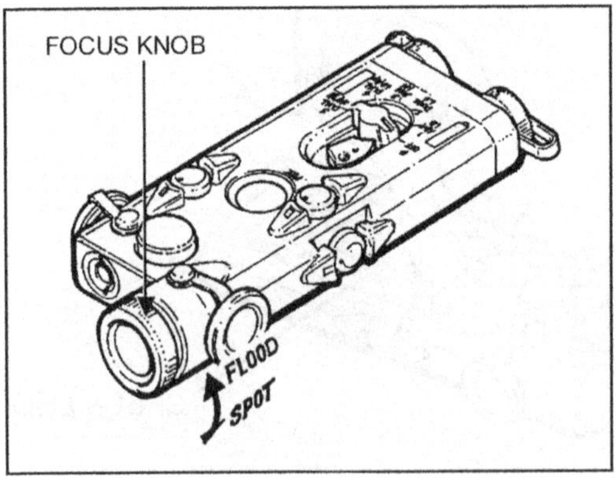

Figure F-11. Using the focus knob.

(7) ***Lens Cap*** (Figure F-12).

(a) The *black lens cap* blocks the AN/PEQ-2A illuminator or aiming laser beam should the device be activated. To use the black lens cap, pull it from its stored location on the side of the AN/PEQ-2A and stretch it over the front of the focus knob or aiming beam so that it fits snugly in place.

(b) The *diffuser lens cap* enables the illuminator or aiming laser to emit in a 45-degree cone (10 feet at 10 feet). To use the diffuser lens cap, pull it from its stored location on the side of the AN/PEQ-2A and stretch it over the front of the focus knob or aiming beam so that it fits snugly in place.

(c) The *neutral density lens cap* enables the AN/PEQ-2A illuminator or aiming laser to be operated in low power. To use the neutral density lens cap, pull it from its stored location on the side of the AN/PEQ-2A and stretch it over the front of the focus knob or aiming beam so that it fits snugly in place.

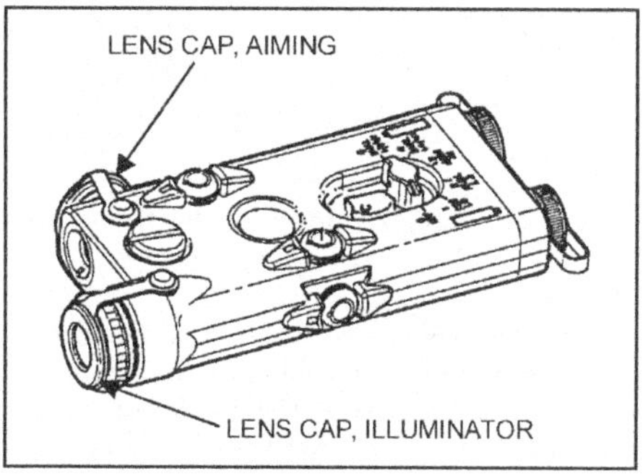

Figure F-12. Installing the lens caps.

(8) ***Boresight Adjusters*** (Figure F-13). The AN/PEQ-2A is equipped with boresight adjusters for zeroing the aiming beam and illumination beam. The AN/PEQ-2A adjusters move the beams in true horizontal and vertical directions. When zeroing the AN/PEQ-2A, it is best to zero the aiming beam to the weapon and then align the illumination beam to the aiming beam.

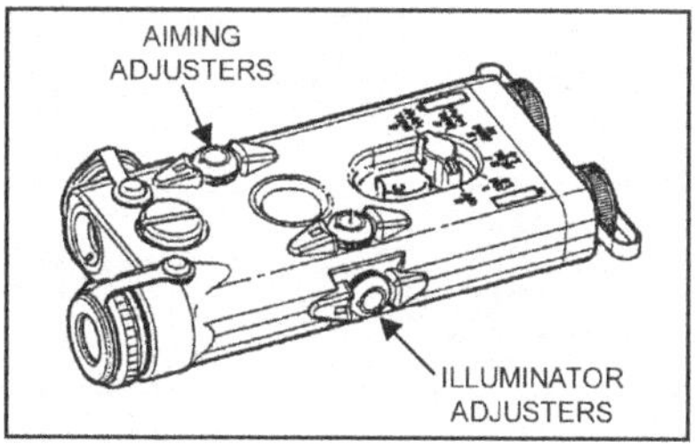

Figure F-13. Boresight adjusters.

F-3. AN/PAS-13 (V2) MEDIUM WEAPON THERMAL SIGHT and AN/PAS-13 (V3) HEAVY WEAPON THERMAL SIGHT

The AN/PAS-13 (V2) medium weapon thermal sight (MWTS) and the AN/PAS-13 (V3) heavy weapon thermal sight (HWTS) (Figure F-14) are silent, lightweight, compact, and durable battery-powered infrared imaging sensors that operate with low battery consumption.

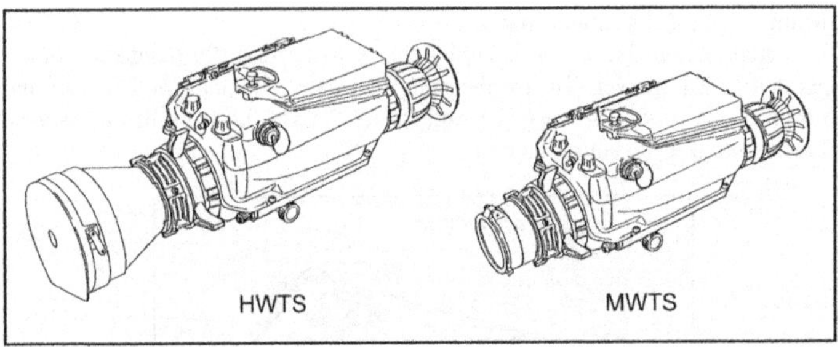

Figure F-14. HWTS and MWTS models of the thermal weapon sight.

a. **Description.**

(1) Both the MWTS and the HWTS are referred to as singular thermal weapon sights (TWS). The TWS is capable of target acquisition under conditions of limited visibility such as darkness, smoke, fog, dust, and haze. The TWS operates effectively at night and can also be used during the daytime. Infrared light is received through the telescope, detected by an IR sensor, converted to digital data, processed, and then displayed for the user. The TWS is composed of two functional groups: the telescope and the basic sensor.

(a) *Telescope.* The telescope receives IR light emitting from an intended target and its surroundings. The telescope magnifies and projects the IR light onto the scanner on the basic sensor (Figure F-15).

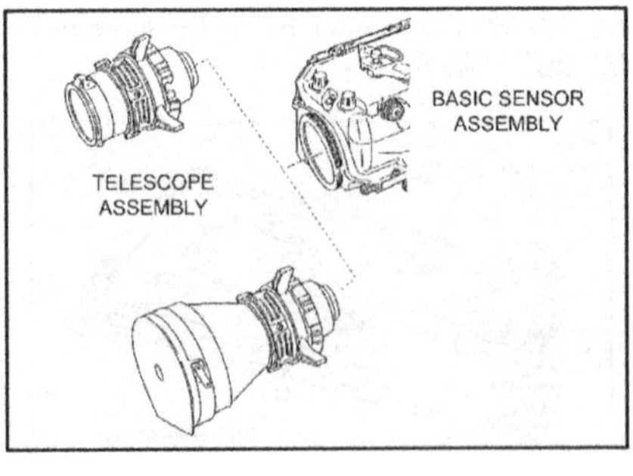

Figure 3-6. TWS configurations.

(b) *Basic sensor.* The scanner reflects the IR light received from the telescope onto the detective assembly. The detective assembly senses the IR light and converts it to video. The sensor's electronics condition the video for display on the LED array. The LED array illuminates the IR image along with the reticle. The light from the LED array is reflected off the scanner to form an image at the eyepiece.

(2) The only difference between the MWTS and the HWTS are the telescopes, which are different magnifications and hold different reticles. The basic sensor on the two models is the same. The MWTS fits the M4 carbine, the M4 modular weapon system, the M16A2, the M16A4 (not shown), the M249, the M60 machine gun, the M240B machine gun (not shown), and the M136 grenade launcher (Figure F-16). The HWTS fits the squad leader's M16, the M24 sniper rifle, the M2, caliber .50 machine gun, and the MK-19 grenade launcher (Figure F-17).

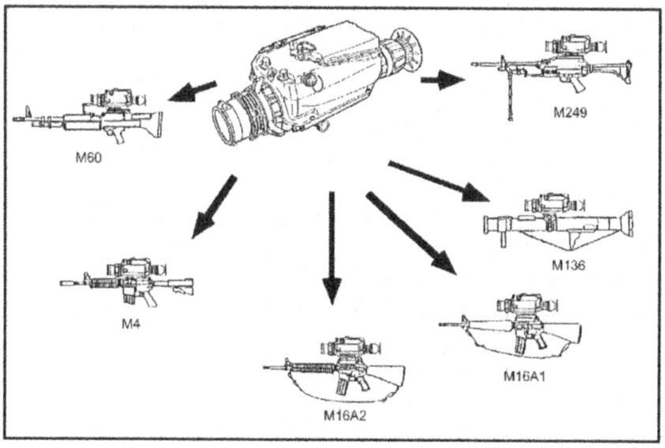

Figure F-16. MTWS.

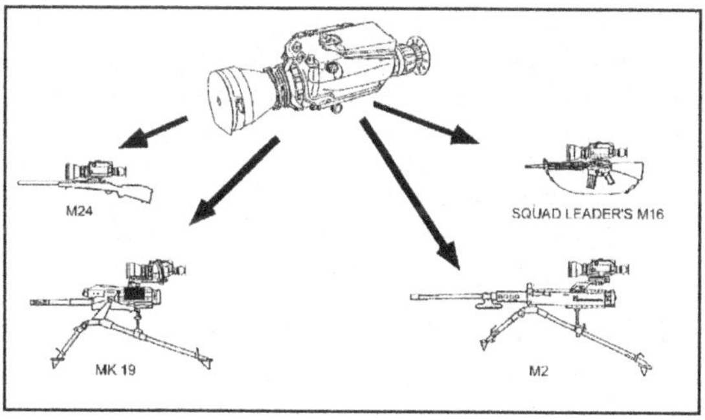

Figure F-17. HTWS.

b. **Technical Data.**

	MWTS	HWTS
Field of view (FOV):		
Narrow	9 degrees	3 degrees
Wide	15 degrees	9 degrees
Telescope magnification:		
Narrow FOV	3.3X	10X
Wide FOV	2X	3.3X
Length:	15.5 inches	18 inches
Width:	6.25 inches	6.25 inches
Height:	6.25 inches	6.25 inches
Weight	4.1 pounds	4.5 pounds
Power consumption:		
ON mode	12.0 watts	12.0 watts
STANDBY mode	5.5 watts	5.5 watts
Battery life:		
71% STANDBY mode	10 hours	10 hours
EMERGENCY mode	3.5 hours	3.5 hours

c. **Operation.**

(1) The TWS has three modes of operation: STANDBY, ON, and EMERGENCY.

(a) STANDBY mode. When the system is first turned on, the TWS begins a cool down period of approximately two minutes. After the cool down period, the TWS enters the STANDBY mode. During the STANDBY mode, power is not applied to the scanner or display in order to extend the life of the battery.

(b) ON mode. When the TWS is in the STANDBY mode and pressure is applied to the eyecup, the TWS switches to the ON mode, and a switch engages to provide power to the scanner and display. After a three-second delay, the system is fully operational.

(c) EMERGENCY mode. When switched to the EMERGENCY mode, the TWS continuously applies power to the entire system. This allows the operator to bypass the three-second delay experienced when switching from the STANDBY to the ON mode. Since power is applied to the entire system while in the emergency mode, battery life is greatly reduced.

(2) Controls and indicators are as shown in Figure 3-18.

(a) The BRIGHTNESS CONTROL is a nine-position rotary switch with an off detent position (turned fully counterclockwise). The purpose is to turn the system on or off and adjust the brightness of the eyepiece display.

(b) The CONTRAST CONTROL adjusts the contrast of the thermal image displayed on the raster. It has an automatic and a manual mode.

(c) The FOCUS RING adjusts the telescope focus from 20 meters to infinity. It requires a manual adjustment and affects both the wide and narrow fields of view.

(d) The FIELD OF VIEW RING is located on the telescope. It has a wide and a narrow field of view. The wide FOV is for using low magnification during target detection, and the narrow FOV is for using high magnification during recognition and engagement.

(e) The RETICLE SELECT SWITCH selects one of the available reticles depending on the TWS model (medium or heavy). It must be held for two seconds to enable reticle changes. After two seconds, release the switch to cycle to the next reticle. This control is disabled after ten seconds of inactivity.

(f) The RETICLE ADJUST SWITCH adjusts the reticle aiming features in azimuth and elevation. It is used during zeroing, and it must be held for two seconds to allow changes to be made. After two seconds, each press moves the reticle aiming features one increment. This control is also disabled after ten seconds of inactivity.

(g) The BLACK/WHITE POLARITY SWITCH selects the polarity of the thermal image displayed on the raster. The initial setting is "white hot." The polarity switch affects the appearance of the target.

(h) The DIOPTER FOCUS RING adjusts the focus of the raster and indicators to the operator's eye. It ranges from +2 to -6 diopters.

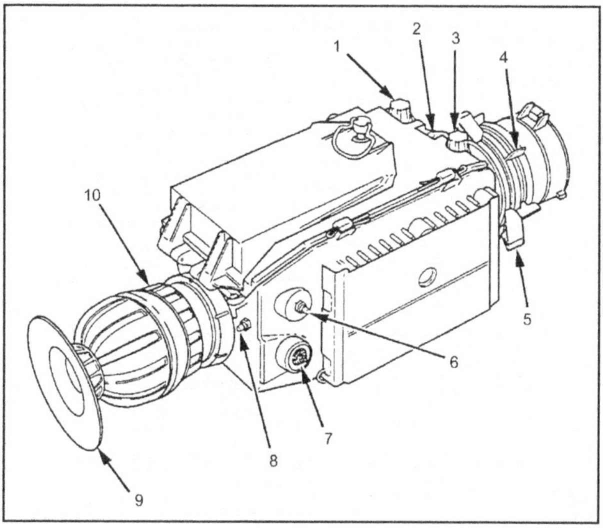

(1) BRIGHTNESS CONTROL	(6) RETICLE SELECT SWITCH
(2) EMERGENCY SWITCH CONTROL	(7) RETICLE ADJUST SWITCH
(3) CONTRAST CONTROL	(8) BLACK/WHITE POLARITY SWITCH
(4) FOCUS RING	(9) EYECUP
(5) FIELD OF VIEW RING	(10) DIOPTER FOCUS RING

Figure F-18. TWS controls and indicators.

(i) The EYEPIECE INDICATORS (Figure F-19) illuminate as follows:

- NOT COOL when the detectors are not cool enough for proper operation.
- WHT HOT/BLK HOT polarity.
- EMER during emergency mode operation.
- LOW when battery power has approximately fifteen minutes of useful power left.

(j) The COOLDOWN PERIOD INDICATORS blink the first ten seconds after the system is turned on. After ten seconds, the NOT COOL indicator is lit and the POLARITY, EMERGENCY, and LOW BATTERY indicators function normally.

(k) The DISPLAY RASTER displays the thermal image with a superimposed reticle.

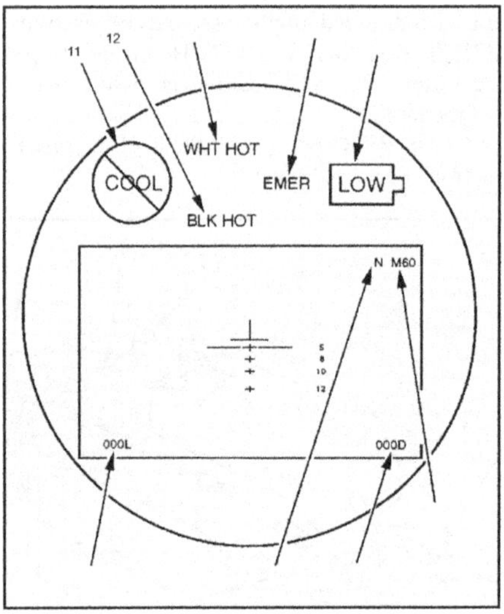

Figure F-19. Eyepiece indicators.

F-4. M136 AT4/M287 AT4 SUBCALIBER TRACER TRAINER SIGHT ALIGNMENT PROCEDURES

The fundamentals of marksmanship are the same as during the daytime with two exceptions. When using aided vision devices, the firer should hold the mounting bracket with the left hand. This method provides greater stability and helps to balance the AT4 against the added weight of the sight. When aiming the AT4, remember to aim by placing the infrared aiming light center mass on the target.

> **WARNING**
> **When firing the M136 AT4, do not place your face within 3 inches of the mounted sight. The recoil of the AT4 could cause injury you or damage to the equipment.**

a. Select a stable position for the weapon.

b. Open the M136 front and rear sight covers. During limited visibility conditions, the 7-millimeter peephole may be necessary for zeroing.

c. Set the rear sight for 200 meters and choose the appropriate scenario.

(1) Place a 25 meter zero target with a predetermined offset at 10 meters and align the weapon's fixed sight to coincide with the 4-centimeter aiming point of the zero target.

(2) Select a suitable target at 200 meters and align the weapon's fixed sight center mass of the target.

d. Turn on the infrared aiming light. Use the AN/PVS-7B or AN/PVS-14 to observe the infrared aiming light (without moving the weapon).

e. Adjust the infrared aiming light so that it coincides with the 2-centimeter offset aiming point of the 25-meter zero target. If the 200-meter method is used, adjust the aiming point of the mounted sight to coincide center mass on the 200-meter target.

f. Repeat the procedure until the aiming light is sight aligned to the weapon by adjusting the controls on the AN/PAQ-4C.

NOTE: Use the same procedures to conduct sight alignment of the AN/PEQ-2.

g. Conduct a preventive maintenance inspection before the instructional and qualification tables are fired. Once the soldiers have completed Tables I and III (day tables), they receive additional instruction on the use of aided vision devices for night operations.

(1) ***Table I (Instructional Day)***. This scenario consists of 12 targets from 100 to 300 meters. The standard to proceed to the next table is 6 out of 12 targets successfully engaged.

(2) ***Table II (Instructional Night)***. This scenario consists of 12 targets from 100 to 300 meters. The standard to proceed to the next table is 6 out of 12 targets successfully engaged at night.

(3) ***Table III (Qualification Day)***. This scenario consists of 10 targets from 100 to 300 meters. The standard to proceed to the next table is 8 out of 10 targets successfully engaged.

(4) ***Table IV (Qualification Night)***. This scenario consists of 10 targets from 100 to 300 meters. The standard to proceed to the next table is 6 out of 10 targets successfully engaged.

h. The night sustainment training strategy should be conducted over a one-day time frame and consists of the following:

(1) ***Night Vision Goggles Familiarization Training***. Soldiers receive instruction on the proper use and fit of night vision goggles, to include characteristics and capabilities, maintenance, and mounting procedures.

(2) ***Infrared Aiming Light Familiarization Training***. Soldiers boresight the weapon and aided vision device at a range of 10 meters.

(3) ***Fundamentals of Firing (Body Position with Head Harness and NVD, Foxhole and Prone)***. Soldiers review and practice firing positions and fundamentals of marksmanship and any changes that may occur by use of the infrared aiming device.

(4) ***Weapon and Equipment Precombat Inspection***. The weapon and infrared aiming device is inspected and properly mounted. Inspect to ensure that the infrared aiming

device is mounted securely to the mount and that the mount is tightly secured to the weapon. A review of clearing or misfire procedures is important.

(5) ***Infrared Aiming Light Boresight to Weapon***. Each soldier boresights the weapon and infrared aiming device at a range of 10 meters. If a bore light is not available, then zero the weapon and sight by using the specified zeroing procedure for that weapon/sight combination.

GLOSSARY

AC	Active Component
ANCOC	advanced noncommissioned officer course
AN/PVS	Army Navy passive vision sight
ARTEP	Army Training and Evaluation Program
backblast	The backward blast created by igniting the propellant in a weapon designed to be as recoilless as possible.
blast	An explosive or violent detonation that produces a direct effect on its target in addition to radiant heat and overpressure at its point of origin.
BMP	a type of non-US fighting vehicle built by the former Soviet Union
BNCOC	basic noncommissioned officer course
BRDM	a type of non-US scout car built by the former Soviet Union
BTR	a type of non-US personnel carrier built by the former Soviet Union
C	Centigrade
CS	combat support
CSS	combat service support
DA	Department of the Army
DKIE	decontamination kit individual equipment
DOD	Department of Defense
DS	direct support
F	Fahrenheit
FHT	field handling trainer

fps	feet per second
FTX	field training exercise
HE	high-explosive
HEAT	high-explosive antitank
HELLFIRE	heliborne laser fire and forget (a missile system)
HMTS	medium weapon thermal sight
HWTS	heavy weapon thermal sight
IAW	in accordance with
IET	initial entry training
IR	infrared
kmph	kilometers per hour
LAW	light antitank weapon
LED	light emitting diode
METL	mission-essential task list
MILES	Multiple Integrated Laser Engagement System
mm	millimeter(s)
MMTT	manned moving target tank
MOPP	mission-oriented protective posture
MOUT	military operations on urbanized terrain
mph	miles per hour
mps	meters per second
MTP	mission training plan

MWTS	medium weapon thermal sight
NA	not applicable
NATO	North Atlantic Treaty Organization
NBC	nuclear, biological, chemical
NCO	noncommissioned officer
NSN	national stock number
NVD	night vision device
OIC	officer in charge
overpressure	increased atmospheric pressure (positive overpressure), followed by a wave of decreased atmospheric pressure (negative overpressure), produced around the origin of an explosive or violent detonation
PLDC	primary leadership development course
POI	program of instruction
promethium	a radioactive metallic element of the rare-earth group used to illuminate range markings on the front sight of the M72-series LAW
RC	Reserve Component
RPG	rocket-propelled grenade (non-US)
STX	situational training exercise
SMCT	soldier's manual of common tasks
SOP	standing operating procedure
spall	small fragment or chip
STP	soldier training publication

STRAC	standards in training commission
TOC	tactical operations center
TP	training-practice
TPIAL	target pointer/illuminator/aiming light, AN/PEQ-2A
TRADOC	(US Army) Training and Doctrine Command
trc	track (as used in STRAC manual)
TRP	target reference point
TSC	training support center
TWS	thermal weapon sight

REFERENCES

SOURCES USED

These are the sources quoted or paraphrased in this manual.

TM 9-1055-886-12&P	Operator and Unit Maintenance Manual for M287 9-MM Tracer Bullet Training Device. 25 October 1989.
TM 9-1315-886-12	Operator's and Unit Maintenance Manual for Launcher and Cartridge, 84 Millimeter: M136 (AT4). 15 May 1990.
TM 9-1340-214-10	Operator's Manual for 66-MM Light Antitank Weapon (LAW) System M72A1, M72A2 with Coupler, M72A3, and Practice Rocket Launcher M190 with M73 Practice Rocket. 20 February 1992.
TM 11-5855-213-10	Operator's Manual for Night Vision Sight, Individual-Served Weapon AN/PVS-4. 1 February 1993.
TM 11-5855-213-23&P	Unit and Direct Support Maintenance Manual for Night Vision Sight, Individual-Served Weapon, AN/PVS-4. 1 June 1993.

DOCUMENTS NEEDED

These documents must be available to the intended users of this publication.

DA Form 7323-R	M72-Series LAW Scorecard. July 1994.
DA Form 7324-R	M136 AT4 Scorecard. July 1994.
DA Pamphlet 350-38	Standards in Weapons Training. 3 July 1997.
*DOD 5100.76-M	Physical Security of Sensitive Conventional Arms, Ammunition, and Explosives. 3 February 1983.
FM 7-7	The Mechanized Infantry Platoon and Squad (APC). 15 March 1985.
FM 7-7J	The Mechanized Infantry Platoon and Squad (Bradley). 7 May 1993.
FM 7-8	The Infantry Platoon and Squad. 1 March 2001.
FM 25-101	Battle Focused Training. 30 September 1990.

M72-SERIES LAW SCORECARD

For use of this form, see FM 23-25; the proponent agency is TRADOC

NAME ______________________ SSN ______________________ DATE __________

GRADE ______ RATING ______ UNIT ______________ WEATHER (circle one) CLEAR RAIN FOG SNOW

DATA REQUIRED BY THE PRIVACY ACT OF 1974

AUTHORITY: 10 USC 3012(g) / Executive Order 9397 **PRINCIPAL PURPOSE(S):** Record individual performance. **ROUTINE USES:** Evaluate individual proficiency and determine proficiency level. SSN is used for positive identification purposes only. **MANDATORY OR VOLUNTARY DISCLOSURE AND EFFECT ON INDIVIDUAL NOT PROVIDING INFORMATION:** Voluntary. Individuals not providing information cannot be rated/scored on a mass basis.

TABLE	TYPE OF FIRE	TYPE OF TARGET	ROUND	RANGE (M)	FIRING POSITION	HIT	MISS
1	Instructional DAY	Stationary	1	125 to 150	Standing	___	___
			2	150 to 175	Kneeling	___	___
			* 3	150 to 200	Sitting	___	___
			* 4	125 to 150	Prone	___	___
					Table Subtotals	___	___

FIRER'S SIGNATURE ______________________ DATE ________ SCORER'S SIGNATURE ______________________ DATE ________

TABLE	TYPE OF FIRE	TYPE OF TARGET	ROUND	RANGE (M)	FIRING POSITION	HIT	MISS
2	Instructional DAY	Moving	1	100 to 200	Standing	___	___
			2	100 to 200	Sitting	___	___
			* 3	100 to 200	Kneeling	___	___
					Table Subtotals	___	___

FIRER'S SIGNATURE ______________________ DATE ________ SCORER'S SIGNATURE ______________________ DATE ________

TABLE	TYPE OF FIRE	TYPE OF TARGET	ROUND	RANGE (M)	FIRING POSITION	HIT	MISS
3	Instructional NIGHT	Stationary	1	100 to 150	Kneeling	___	___
			2	100 to 150	Sitting	___	___
			* 3	100 to 150	Sitting	___	___
					Table Subtotals	___	___

FIRER'S SIGNATURE ______________________ DATE ________ SCORER'S SIGNATURE ______________________ DATE ________

* Firer must wear a protective mask when firing this round.

CONSOLIDATED SCORES

FIRER'S SIGNATURE ______________________ DATE __________ TOTAL MISSES ______

SCORER'S SIGNATURE ______________________ DATE __________ RANK ______

RANGE OIC'S SIGNATURE ______________________ DATE __________ RANK ______

INSTRUCTOR NOTES	
ALL FIRING TABLES	• Do not announce the ranges to targets. • Fire all engagements in sequence.

DA Form 7323-R, JUL 94

M136 AT4 SCORECARD

For use of this form, see FM 23-25; the proponent agency is TRADOC

NAME ______________________ SSN ______________________ DATE __________

GRADE ______ RATING ______ UNIT ______________ WEATHER (circle one) CLEAR RAIN FOG SNOW

DATA REQUIRED BY THE PRIVACY ACT OF 1974

AUTHORITY: 10 USC 3012(g) / Executive Order 9397. **PRINCIPAL PURPOSE(S):** Record individual performance. **ROUTINE USES:** Evaluate individual proficiency and determine proficiency level. SSN is used for positive identification purposes only. **MANDATORY OR VOLUNTARY DISCLOSURE AND EFFECT ON INDIVIDUAL NOT PROVIDING INFORMATION:** Voluntary. Individuals not providing information cannot be rated/scored on a mass basis.

TABLE	TYPE OF FIRE	TYPE OF TARGET	ROUND	RANGE (M)	FIRING POSITION	HIT	MISS
		Stationary	1	150 to 200	Kneeling	___	___
		Stationary	* 2	150 to 200	Kneeling	___	___
		Stationary	3	250 to 300	Sitting	___	___
		Stationary	4	250 to 300	Sitting	___	___
1	Instructional Day	Stationary	5	125 to 150	Prone	___	___
		Stationary	6	125 to 150	Prone	___	___
		MMTT/MAS	7	150 to 200	Kneeling	___	___
		MMTT/MAS	* 8	150 to 200	Kneeling	___	___
		MMTT/MAS	9	200 to 250	Standing	___	___
		MMTT/MAS	*10	200 to 250	Fighting position, unsupported	___	___
		MMTT/MAS	11	250 to 300	Kneeling	___	___
		MMTT/MAS	12	250 to 300	Kneeling	___	___
					Table Subtotals	___	___

FIRER'S SIGNATURE ______________________ DATE ________ SCORER'S SIGNATURE ______________________ DATE ________

TABLE	TYPE OF FIRE	TYPE OF TARGET	ROUND	RANGE (M)	FIRING POSITION	HIT	MISS
		Stationary	1	125 to 150	Prone	___	___
		Stationary	2	125 to 150	Prone	___	___
		Stationary	* 3	150 to 200	Sitting	___	___
		Stationary	4	150 to 200	Sitting	___	___
		Stationary	5	250 to 300	Kneeling	___	___
2	Instructional Night	Stationary	6	250 to 300	Kneeling	___	___
		MMTT/MAS	7	125 to 150	Standing	___	___
		MMTT/MAS	8	125 to 150	Standing	___	___
		MMTT/MAS	* 9	200 to 250	Kneeling	___	___
		MMTT/MAS	10	200 to 250	Kneeling	___	___
		MMTT/MAS	11	150 to 200	Standing	___	___
		MMTT/MAS	12	150 to 200	Fighting position, unsupported	___	___
					Table Subtotals	___	___

FIRER'S SIGNATURE ______________________ DATE ________ SCORER'S SIGNATURE ______________________ DATE ________

* Firer must wear a protective mask when firing this round.

INSTRUCTOR NOTES	
ALL FIRING TABLES	• Do not announce the ranges to targets. • Fire all engagements in sequence. • Fire Table 1 before Table 3; fire Table 2 before Table 4.
FIRING TABLES 2 THROUGH 5	• Ensure no one fires at front or rear of a manned moving tank target. • Total hits and misses, and enter results in the consolidated scores block on the back of this scorecard.

INDEX

Made in the USA
Monee, IL
04 November 2026

67386426R00075